SciPy Recipes

A cookbook with over 110 proven recipes for performing
mathematical and scientific computations

L. Felipe Martins
Ruben Oliva Ramos
V Kishore Ayyadevara

BIRMINGHAM - MUMBAI

SciPy Recipes

First published: December 2017

Production reference: 1191217

Published by Packt Publishing Ltd.
Livery Place
35 Livery Street
Birmingham
B3 2PB, UK.
ISBN 978-1-78829-146-0

www.packtpub.com

Credits

Authors
L. Felipe Martins
Ruben Oliva Ramos
V Kishore Ayyadevara

Reviewer
Juan Tomás Oliva Ramos

Commissioning Editor
Amey Varangaonkar

Acquisition Editor
Tushar Gupta

Content Development Editor
Aishwarya Pandere

Technical Editor
Prasad Ramesh

Copy Editor
Safis Editing

Project Coordinator
Nidhi Joshi

Proofreader
Safis Editing

Indexer
Pratik Shirodkar

Graphics
Tania Dutta

Production Coordinator
Arvindkumar Gupta

About the Authors

L. Felipe Martins has a PhD in applied mathematics from Brown University and is currently an associate professor in the Department of Mathematics at Cleveland State University. His main research areas are applied probability and scientific computing. He has taught applied mathematics courses at all levels, including linear algebra, differential equations, probability, and optimization, and uses Python as an instructional tool in all courses. He is the author of two books, *IPython Notebook Essentials* and *Mastering Python Data Analysis*.

Ruben Oliva Ramos is a computer systems engineer from Tecnologico de Leon Institute, with a master's degree in computer and electronic systems engineering and a specialization in teleinformatics and networking from the University of Salle Bajio in Leon, Guanajuato, Mexico. He has more than 5 years of experience of developing web applications to control and monitor devices connected with Arduino and Raspberry Pi, using web frameworks and cloud services to build the Internet of Things applications.

He is a mechatronics teacher at the University of Salle Bajio and teaches students of the master's degree in design and engineering of mechatronics systems. Ruben also works at Centro de Bachillerato Tecnologico Industrial 225 in Leon, Guanajuato, Mexico, teaching subjects such as electronics, robotics and control, automation, and microcontrollers on the Mechatronics Technician career course; he is a consultant and developer for projects in areas such as monitoring systems and datalogger data using technologies (such as Android, iOS, Windows Phone, HTML5, PHP, CSS, Ajax, JavaScript, Angular, and ASP.NET), databases (such as SQLite, MongoDB, and MySQL), web servers (such as Node.js and IIS), hardware programming (such as Arduino, Raspberry Pi, Ethernet Shield, GPS, GSM/GPRS, and ESP8266), and control and monitor systems for data acquisition and programming.

Ruben is the author of the following books by Packt: *Internet of Things Programming with JavaScript*, *Advanced Analytics with R and Tableau*, and *Raspberry Pi 3 Home Automation Projects*.

He is also involved in monitoring, controlling, and acquiring of data with Arduino and Visual Basic .NET for Alfaomega.

I would like to thank my savior and lord, Jesus Christ, for giving me the strength and courage to pursue this project; to my dearest wife, Mayte; our two lovely sons, Ruben and Dario; my dear father, Ruben; my dearest mom, Rosalia; my brother, Juan Tomas; and my sister, Rosalia, whom I love, for all their support while reviewing this book, for allowing me to pursue my dream and tolerating me not being with them after my busy days.
I'm very grateful to Packt Publishing for giving me the opportunity to collaborate as an author and reviewer, and to belong to this honest and professional team.

V Kishore Ayyadevara has over 9 years of experience of using analytics to solve business problems and setting up analytical work streams through his work at American Express, Amazon, and more recently a retail analytics consulting start-up. He is an MBA graduate from IIM Calcutta and also an electronics and communications engineer.

He worked in the fields of credit risk analytics, supply chain analytics, and consulting for multiple FMCG companies to identify ways to improve their profitability.

His interests lie in translating a business problem into a data-related problem by demystifying complexity in data science and identifying ways to further embed analytics in business.

I would like to dedicate my work on this book to my dear parents—Hema and Subrahmanyeswara Rao; my lovely wife—Sindhura and my dearest daughter, Hemanvi. This work would not have been possible without their support and encouragement.

About the Reviewer

Juan Tomás Oliva Ramos is an environmental engineer from the University of Guanajuato, Mexico, with a master's degree in administrative engineering and quality. He has more than 5 years of experience in managing and developing patents, technological innovation projects, and developing technological solutions through statistical control of processes.

He has been a teacher of statistics, entrepreneurship, and technological development of projects since 2011. He has always maintained an interest in improvement and innovation in processes through technology. He became an entrepreneur mentor and technology management consultant. Juan started a new department of technology management and entrepreneurship at Instituto Tecnologico Superior de Purisima del Rincon, Mexico.

He has worked on the book *Wearable Designs for Smart Watches, Smart TVs, and Android Mobile Devices.*

Juan has also developed prototypes using programming and automation technologies to improve operations, which have been registered for patents.

I want to thank my beautiful wife, Brenda, our two magic princesses (Maria Regina and Maria Renata) and our next member (Angel Tadeo), despite the distance they are always in my heart and in my mind.
I thank Packt and Nidhi Joshi for giving me the opportunity to review this amazing book and to collaborate with a group of committed people.

www.PacktPub.com

For support files and downloads related to your book, please visit www.PacktPub.com. Did you know that Packt offers eBook versions of every book published, with PDF and ePub files available? You can upgrade to the eBook version at www.PacktPub.com and as a print book customer, you are entitled to a discount on the eBook copy.

Get in touch with us at service@packtpub.com for more details. At www.PacktPub.com, you can also read a collection of free technical articles, sign up for a range of free newsletters and receive exclusive discounts and offers on Packt books and eBooks.

https://www.packtpub.com/mapt

Get the most in-demand software skills with Mapt. Mapt gives you full access to all Packt books and video courses, as well as industry-leading tools to help you plan your personal development and advance your career.

Why subscribe?

- Fully searchable across every book published by Packt
- Copy and paste, print, and bookmark content
- On demand and accessible via a web browser

Customer Feedback

Thanks for purchasing this Packt book. At Packt, quality is at the heart of our editorial process. To help us improve, please leave us an honest review on this book's Amazon page at https://www.amazon.com/dp/1788291468.

If you'd like to join our team of regular reviewers, you can email us at customerreviews@packtpub.com. We award our regular reviewers with free eBooks and videos in exchange for their valuable feedback. Help us be relentless in improving our products!

Table of Contents

Preface

The SciPy stack is a collection of open source libraries of the powerful Python scripting language, together with its interactive shells. This environment offers a cutting-edge platform for numerical computation, programming, visualization, and publishing, and is used by some of the world's leading mathematicians, scientists, and engineers. It works on any operating system that supports Python and is very easy to install and completely free of charge! It can effectively transform into a data-processing and system-prototyping environment.

The SciPy stack is a popular Python ecosystem used for mathematical and scientific computing tasks. It can be used to perform a variety of data science tasks, from data manipulation to visualization. Utilizing the offerings of SciPy to perform your data science tasks is a very tricky proposition.

This book will show you how you can put to use the various functionalities offered by the SciPy stack in the most efficient way possible. With the help of this book, you will solve real-world problems in linear algebra, numerical analysis, visualization, and much more, including independent recipes drawn from the fields of statistics, scientific computation, and visualization. You will master the different tasks associated with using SciPy and its related libraries, such as NumPy, Matplotlib, pandas and more, in the best way. This book will ensure that you have a practical understanding of not only how a particular feature in SciPy stack works but also its applications in real-world problems.

What this book covers

Chapter 1, *Getting to Know the Tools*, explains how to install and configure all the packages required to set up and configure an environment dedicated to scientific computing in Python. The chapter considers several different setup options in the three main operating systems available to users: Windows, macOS, and Linux.

Chapter 2, *Getting Started with NumPy*, presents the essential recipes for efficient use of NumPy, the Python package for numerical computations on which SciPy is based.

Chapter 3, *Using Matplotlib to Create Graphs*, is a thorough discussion of Matplotlib, the plotting library included with NumPy and SciPy, concentrating on the skills required to display the results of technical computations.

Chapter 4, *Data Wrangling with pandas*, shows how to use pandas, a powerful package for data handling and analysis in Python.

Chapter 5, *Matrices and Linear Algebra*, covers performing the various matrix data manipulation techniques such as basic matrix operations, solving linear systems, finding eigenvalues and eigenvectors, calculating the singular value decomposition, and sparse matrix manipulation techniques that are potentially used in recommender systems using SciPy.

Chapter 6, *Solving Equations and Optimization*, discusses the solutions of numerical equations and systems of equations, as well as the solution of maximization/minimization problems.

Chapter 7, *Constants and Special Functions*, presents the numerical constants and special functions that are available in SciPy.

Chapter 8, *Calculus, Interpolation, and Differential Equations*, shows how to solve essential calculus problems, including integration, differentiation, interpolation, and differential equations.

Chapter 9, *Statistics and Probability*, covers the various statistics and probability measures such as PMF, PDF, CDF, and multivariate Gaussian distributions using SciPy.

Chapter 10, *Advanced Computations with SciPy*, discusses the advanced computations available in SciPy that are of a more specific nature.

What you need for this book

To get the most from this book, the reader needs to know the basics of Python; it's not necessary that the reader has the ability to program because the first chapter explains how to install the plugins needed to work with SciPy. The following are the software and OS requirements:

- SciPy 1.0.0
- NumPy v1.13
- Matplotlib 2.1.0
- Python 2.7, 3.5, and 3.6
- Python Data Analysis Library (v0.21.0)
- SymPy 1.1.1
- Released libraries
- OS required: Windows, Mac, or Linux

Who this book is for

Python developers, aspiring data scientists, and analysts who want to get started with scientific computing using Python will find this book to be a useful resource. If you want to learn how to manipulate and visualize your data using the SciPy stack, this book will also help you. A basic understanding of Python programming is all you need to get started.

This book is for readers who want learn more about SciPy in specific topics and gain the basic knowledge required to solve problems. The following are the objectives:

- Tackle sophisticated problems in scientific computing with the SciPy stack
- Get a solid foundation in scientific computing with Python and open source software
- Present common tasks related to SciPy and associated libraries such as NumPy, pandas, and Matplotlib
- Perform mathematical operations and work with the statistical and probability functions in SciPy
- Empower users to further explore the library and find solutions to their own computational needs
- Discuss best practices and efficient methods in the solution of computational problems

Conventions

In this book, you will find a number of text styles that distinguish between different kinds of information. Here are some examples of these styles and an explanation of their meaning. Code words in text, database table names, folder names, filenames, file extensions, pathnames, dummy URLs, user input, and Twitter handles are shown as follows: "There are several ways to create objects of ndarray type".

A block of code is set as follows:

```
x = np.array([[1,2,3,4],
              [5,6,7,8]])
y = np.repeat(x, [1,0,1,0], axis=1)
```

Any command-line input or output is written as follows:

```
pip3 install molecule
```

New terms and **important words** are shown in bold. Words that you see on the screen, for example, in menus or dialog boxes, appear in the text like this: "To create a new notebook, click the **New** button in the top right and select **Python 3** from the menu."

Warnings or important notes appear like this.

Tips and tricks appear like this.

Reader feedback

Feedback from our readers is always welcome. Let us know what you think about this book-what you liked or disliked. Reader feedback is important for us as it helps us develop titles that you will really get the most out of. To send us general feedback, simply email feedback@packtpub.com, and mention the book's title in the subject of your message. If there is a topic that you have expertise in and you are interested in either writing or contributing to a book, see our author guide at www.packtpub.com/authors.

Customer support

Now that you are the proud owner of a Packt book, we have a number of things to help you to get the most from your purchase.

Downloading the example code

You can download the example code files for this book from your account at http://www.packtpub.com. If you purchased this book elsewhere, you can visit http://www.packtpub.com/support and register to have the files emailed directly to you. You can download the code files by following these steps:

1. Log in or register to our website using your email address and password.
2. Hover the mouse pointer on the **SUPPORT** tab at the top.
3. Click on **Code Downloads & Errata**.

4. Enter the name of the book in the **Search** box.
5. Select the book for which you're looking to download the code files.
6. Choose from the drop-down menu where you purchased this book from.
7. Click on **Code Download**.

Once the file is downloaded, please make sure that you unzip or extract the folder using the latest version of:

- WinRAR / 7-Zip for Windows
- Zipeg / iZip / UnRarX for Mac
- 7-Zip / PeaZip for Linux

The code bundle for the book is also hosted on GitHub at `https://github.com/PacktPublishing/SciPy-Recipes`. We also have other code bundles from our rich catalog of books and videos available at `https://github.com/PacktPublishing/`. Check them out!

Errata

Although we have taken every care to ensure the accuracy of our content, mistakes do happen. If you find a mistake in one of our books-maybe a mistake in the text or the code- we would be grateful if you could report this to us. By doing so, you can save other readers from frustration and help us improve subsequent versions of this book. If you find any errata, please report them by visiting `http://www.packtpub.com/submit-errata`, selecting your book, clicking on the **Errata Submission Form** link, and entering the details of your errata. Once your errata are verified, your submission will be accepted and the errata will be uploaded to our website or added to any list of existing errata under the Errata section of that title. To view the previously submitted errata, go to `https://www.packtpub.com/books/content/support` and enter the name of the book in the search field. The required information will appear under the **Errata** section.

Piracy

Piracy of copyrighted material on the internet is an ongoing problem across all media. At Packt, we take the protection of our copyright and licenses very seriously. If you come across any illegal copies of our works in any form on the internet, please provide us with the location address or website name immediately so that we can pursue a remedy. Please contact us at `copyright@packtpub.com` with a link to the suspected pirated material. We appreciate your help in protecting our authors and our ability to bring you valuable content.

Questions

If you have a problem with any aspect of this book, you can contact us at
questions@packtpub.com, and we will do our best to address the problem.

1
Getting to Know the Tools

In this chapter, we cover the following recipes:

- Installing Anaconda on Windows
- Installing Anaconda on macOS
- Installing Anaconda on Linux
- Checking the Anaconda installation
- Installing SciPy from a binary distribution on Windows
- Installing SciPy from a binary distribution on macOS
- Installing SciPy from source on Linux
- Installing optional packages with `conda`
- Installing packages with `pip`
- Setting up a virtual environment with `conda`
- Creating a virtual environment for development with `conda`
- Creating a `conda` environment with a different version of a package
- Using `conda` environments to run different versions of Python
- Creating Virtual environments with venv
- Running SciPy in a script
- Running SciPy in Jupyter
- Running SciPy in Spyder
- Running SciPy in PyCharm

Introduction

In this chapter, we discuss the available options for setting up and running the SciPy stack and associated tools. We present solutions for all the major platforms and consider different scenarios. Readers are advised to browse through all installation options before deciding which option fits their workflow and computational needs. After reading this chapter, the reader will understand all different options for setting up a full-fledged environment in Python for computational and data science.

The recipes in this chapter assume the use of the following tools:

- The Command Prompt, also known as Terminal in Linux and macOS. Each operating system has a different way of accessing the default Command Prompt:
    - In Windows, open the search bar and type `cmd`.
    - In macOS, the Terminal app is in the **Applications-Utilities** folder.
    - In Linux, the Command Prompt may be called **xterm** or **Terminal**. In Ubuntu, it can also be started by pressing *Ctrl + Alt + T*.
- A text editor. Sublime Text is a popular multi-platform programmer's editor with many nice features, available at:
  `https://www.sublimetext.com`. Sublime Text is commercial software, but a trial version is available.
  Alternatives available for each operating system are as follows:
    - **Windows**: Notepad is pre-installed on Windows. A free Notepad alternative that adds nice features for free is Notepad++, which can be downloaded from `https://notepad-plus-plus.org`.
    - **macOS**: TextEdit is pre-installed and can be found in the **Applications** folder. An alternative is `nano`, a simple text editor that can be started from a Terminal window.
    - **Linux**: Usually ships with at least one of the following: `gedit`, `nano`, or `vim`, all of which can all be launched from a Terminal window.

If you decide to use the pre-assembled Anaconda distribution, you will also need to download it from the following site: `https://www.continuum.io/downloads`.

Choose the latest 64-bit Python 3 distribution, unless you have an older computer with a 32-bit architecture.

Installing Anaconda on Windows

In this recipe, we will show you how to install Anaconda on a Windows system.

How to do it...

1. Double-click the downloaded .exe installer
2. Accept the software license
3. When prompted for the kind of installation you want, select **Just Me**, and then click **Next**
4. Accept the default installation folder
5. In the next option box, select both **Add Anaconda to my PATH environment variable** and **Register Anaconda as my default Python 3.x**
6. Click **Install** to finish the installation
7. Anaconda will be installed in the C:\Users*username*\Anaconda3 folder
8. Optionally, proceed to the *Checking the Anaconda installation*

Installing Anaconda on macOS

In this recipe, we will show you how to install Anaconda on a macOS system.

How to do it...

1. Double-click the downloaded installer file, which is a file with a .pkg extension
2. Click the **Continue** button to view **ReadMe** and accept the software license
3. When prompted, select the **Install for me only** option and click **Continue**
4. Review the installation options and click the **Install** button
5. Wait until the installation finishes, and then click the **Close** button to quit the installer
6. Anaconda will be installed in the anaconda subfolder of your home folder
7. Optionally, proceed to the *Checking the Anaconda installation* recipe

Installing Anaconda on Linux

In this recipe, we will show you how to install Anaconda on a Linux system.

How to do it...

1. Open a Terminal window on the folder containing the installer and run the following command, replacing the version number, x.x.x, with the corresponding value for file you downloaded:

   ```
   bash Anaconda3-x.x.x-Linux-x86_64.sh
   ```

2. Review the license agreement and accept it. Enter yes when prompted to continue the installation
3. When asked if you want to prepend the Anaconda3 installation location to your path, answer yes
4. Wait until the installer stops
5. Anaconda will be installed in the anaconda3 subfolder of your home folder.
6. Optionally, proceed to the *Checking the Installation* recipe

Checking the Anaconda installation

This recipe shows you how to do some basic checking. We will verify that the software can be started and that the correct version is being used.

How to do it...

1. Open a new Terminal window and run the following command:

   ```
   python3
   ```

2. Verify the information displayed in the Terminal. It will look like the following:

```
Python 3.6.0 |Anaconda 4.3.1 (x86_64)| (default, Dec 23 2016,
13:19:00)
[GCC 4.2.1 Compatible Apple LLVM 6.0 (clang-600.0.57)] on darwin
Type "help", "copyright", "credits" or "license" for more
information.
>>>
```

3. Notice that, in the output, we can check that we are indeed running the Python 3 version distributed with Anaconda.

You have now successfully installed Anaconda on your computer.

Installing SciPy from a binary distribution on Windows

Windows does not ship with any version of Python pre-installed, which actually makes things easier when we want to install our own version of Python.

On the other hand, the installation of a full SciPy stack in Windows is somewhat more complex, due to conflicts that exist between the Python distribution and certain Windows libraries. We indicate an installation route that has been tested several times, but some trial and error may be necessary due to changes in the distribution.

How to do it...

To make the instructions easier to follow, the installation procedure is broken down into two stages:

- Installing Python
- Install the SciPy stack

Installing Python

1. Go to `https://www.python.org` and download the Python 3 binary distribution for Windows
2. Once the download finishes, double-click on the installation file to start the setup
3. Check the box **Add Python 3.x to PATH**
4. Click the **Install Now** option
5. Select the **Disable the path length limit** option, if available, on the last installation screen
6. Close the installation screen

These steps will install Python in the folder:

`C:\Users\username\Appdata\Local\Programs\Python\Python3x`

To test the installation, start a Command Prompt window and enter the following command:

`python3`

If all is correct, the Python command-line interpreter will start and display information about the version of Python being run. For now, just exit the interpreter by entering, at the >>> Python prompt, the following statement:

`quit()`

Now, let's check if `pip` was correctly installed. Enter the following at the command line:

`pip --version`

This should print information about the currently installed version of `pip3`, including the location where packages will be installed. As long as no errors are reported, the installation is correct.

Installing the SciPy stack

To install SciPy, we need to first download the versions of the library that have been built specifically for Windows. They can be found at the following site: `http://www.lfd.uci.edu/~gohlke/pythonlibs/`.

This page contains a long list of pre-compiled Python packages for Windows. Search the page for `numpy-mkl` and `scipy` and look for a package that matches your operating system and Python distribution. In my case, I found the following two files:

```
numpy-1.12/1+mkl-cp36-cp36m-win_amd64.whl
scipy-0.19.0-cp36-cpm36m-win_amd64.whl
```

Notice that the package names refer to version 3.6 and a 64-bit architecture. Make sure the versions you download match your Python 3 distributions. Open a command window on the directory where the files were saved and enter the following two commands, in the following order:

```
pip install numpy-1.12/1+mkl-cp36-cp36m-win_amd64.whl
pip install scipy-0.19.0-cp36-cpm36m-win_amd64.whl
```

After installing NumPy and SciPy, `pip` can be used to install the other packages directly by running the commands shown as follows:

```
pip install matplotlib
pip install ipython jupyter
pip install pandas sympy nose
```

Let's now test the installation. First, start Python 3 and execute the following statements at the >>> Python prompt:

```
import numpy
import scipy
import matplotlib
import pandas
import IPython
import sympy
```

If you can run all these commands and there are no errors, the installation of the packages is correct. Exit the Python shell by running the following statement:

```
quit()
```

Now, back in the command window, run the following command:

```
ipython
```

This will start IPython and display information about the installed version. For now, simply exit IPython by running the following at the prompt:

```
quit()
```

Finally, let's test the Jupyter Notebook. At the command line, run the following command:

```
jupyter notebook
```

If all is correct, this will start the Jupyter Notebook in your browser after a few seconds. This finishes the installation of Python and the SciPy stack in Windows.

Installing SciPy from a binary distribution on macOS

macOS currently ships with version 2.7 of Python pre-installed. In this recipe, we will show you how to install Python 3 on the Mac without making changes to the original Python distribution. The easiest way to achieve this is to use Homebrew, a package manager for macOS.

How to do it...

The full installation instructions are broken down into the following stages:

- Installing the Xcode command-line tools
- Installing Homewbrew
- Installing Python 3
- Installing the SciPy stack

Installing the Xcode command-line tools

Xcode is the free development environment for macOS distributed by Apple. If you already have Xcode installed on your computer, you can skip this step. If you don't, open a Terminal window and run the following command:

```
xcode-select --install
```

If you get an error message, then the command-line tools are already installed and you can go to the next step. Otherwise, a window will pop up asking for confirmation. Press the **Install** button and, when prompted, accept the license agreement.

To check that the command-line tools were correctly installed, run the following command in the Terminal:

```
gcc -v
```

This command prints information about the `gcc` compiler present in your computer, which will be similar to the output shown as follows:

```
Configured with: --prefix=/Library/Developer/CommandLineTools/usr --with-
gxx-include-dir=/usr/include/c++/4.2.1
Apple LLVM version 8.1.0 (clang-802.0.38)
Target: x86_64-apple-darwin16.4.0
Thread model: posix
InstalledDir: /Library/Developer/CommandLineTools/usr/bin
```

If you get no error message, the command-line tools are properly installed.

Installing Homebrew

Homebrew is a package manager for macOS that makes it easier to install and remove software packages without interfering with system software that ships with the computer. It installs package files to the `/usr/local` directory and makes no changes to system folders. Although it is possible to install Python on the Mac from the source, using Homebrew considerably simplifies the setup process.

To install Homebrew, open a Terminal window and run the following command. Please note that the whole command should be typed in a single Terminal line:

```
/usr/bin/ruby -e "$(curl -fsSL
https://raw.githubusercontent.com/Homebrew/install/master/install)"
```

Follow the on-screen instructions and confirm that you want to install Homebrew. Enter the administrative password for your computer if prompted. On a personal computer, this is usually the same as your login password.

To check that Homebrew was successfully installed, run the following command:

```
brew -v
```

This command outputs information about the current Homebrew installation, which looks like the following example:

```
Homebrew 1.1.13
Homebrew/homebrew-core (git revision c80e; last commit 2017-04-26)
```

If you get a similar message and no errors, Homebrew is properly installed.

Installing Python 3

Once Homebrew is set up, install Python 3 by running the following command from a Terminal window:

```
brew install python3
```

The installation process will start and may take a few minutes. When it is finished, run the following from the command line:

```
python3
```

If the installation is correct, this will print information about the Python interpreter, shown as follows:

```
Python 3.x.x (default, Apr  4 2017, 09:40:21)
[GCC 4.2.1 Compatible Apple LLVM 8.1.0 (clang-802.0.38)] on darwin
Type "help", "copyright", "credits" or "license" for more information.
>>>
```

You can check that you are indeed running the Python distribution that you installed by checking the version number, indicated by 3.x.x in the preceding sample. You can now exit Python by running the following command at the >>> Python prompt:

```
quit()
```

We are going to use the pip3 Python package manager to install the SciPy stack. To check that pip3 was correctly installed, run the following statement from the command line:

```
pip3 --version
```

This will print to the Terminal information about the currently installed version of pip, as shown in the following example:

```
pip 9.0.1 from /home/fmartins/python3/lib/python3.x/site-packages (python
3.x)
```

Verify that you are indeed running the version of `pip3` associated with your installation of Python by checking the version number, indicated by `3.x` in the preceding sample output. If no error message is issued, the setup was completed correctly, and you can proceed to install SciPy.

Installing the SciPy stack

To install the SciPy stack, execute each of the following commands on a Terminal window:

```
pip3 install --user numpy scipy matplotlib
pip3 install --user ipython jupyter
pip3 install --user pandas sympy nose
```

We now need to adjust the `PATH` variable in the `.bash_profile` file. Notice that you might not have a `.bash_profile` yet. If you do, it is important to make a backup copy of it by running the following commands at the command line:

```
cd
cp .bash_profile .bash_profile.bak
```

If you get a message stating that `.bash_profile` does not exits, do not worry. We will create one now.

Start your text editor and open the `.bash_profile` file. For example, to use `nano`, a simple text editor included with macOS, run the following in a Terminal window:

```
cd
nano .bash_profile
```

This will create `.bash_profile` if it still does not exist. Add the following line at the end of file, where you need to replace `3.x` by the version of Python you have installed:

```
export PATH="$HOME/Library/Python/3.x/bin:$PATH"
```

Save the file, close the editor, and run the following command from the Terminal window:

```
source .bash_profile
```

This completes the installation of a basic SciPy stack. To test the setup, start Python 3 in the Terminal window by running the following command:

```
python3
```

To check that all packages we need were installed, execute the following lines at the >>> Python prompt. Notice that there will be no output, and, as long as there are no errors, the installation is correct:

```
import numpy
import scipy
import matplotlib
import IPython
import pandas
import sympy
import nose
```

You can now exit the Python shell by running the following statement at the prompt:

```
quit()
```

Let's now check that IPython is accessible from the command line. Run the following line from the Terminal window:

```
ipython
```

This will start IPython, an alternative Python shell with many added features that is required to run Jupyter. A message similar to the following will be printed:

```
Python 3.x.x (default, Apr  4 2017, 09:40:21)
Type 'copyright', 'credits' or 'license' for more information
IPython 6.0.0 -- An enhanced Interactive Python. Type '?' for help.
In [1]:
```

If you get an error when trying to start IPython, the PATH variable was probably not set correctly in the .bash_profile file. You can check the current value of the PATH variable by entering the echo $PATH command. If the path is not correct, edit .bash_profile as indicated previously to correct the error.

Exit Python by entering, the following command in the IPython prompt:

```
quit()
```

Let's now test if we can run the Jupyter Notebook. Run the following command from the Terminal window:

```
jupyter notebook
```

If all is correct, you will see a series of startup messages in the Terminal, and the Jupyter Notebook will eventually start on your browser. If you get an error message, you probably have an incorrectly configured `PATH` variable. Check the preceding tip box for instructions on how to fix it.

This finishes the installation of Python and the SciPy stack in macOS. Please proceed to the *Setting up a virtual environment* section.

Installing SciPy from source on Linux

Since Linux is an umbrella name for a number of distinct operating system configurations, there is no binary distribution that fits all possible Linux flavors.

All modern distributions of Linux come with Python pre-installed. In this recipe, we describe a setup procedure for a local Python 3 installation from source that works on two of the most popular Linux distributions, Ubuntu and Debian. The installation will be located in the user's home directory, and will not conflict with any pre-installed version of Python that exists in the system.

How to do it...

The installation procedure is broken down in the following stages:

- Installing Python 3
- Installing the SciPy Stack

Installing Python 3

Start by opening a Terminal window and running the following commands, one at a time. You will be required to enter the administrative password for your system which, on a personal computer, is usually your own login password:

```
sudo apt-get install build-essential
sudo apt-get install sqlite3 libsqlite3-dev
sudo apt-get install bzip2 libbz2-dev
sudo apt-get install libreadline-dev libncurses5-dev
sudo apt-get install libssl-dev tcl-dev tk-dev python3-tk
sudo apt-get install zlib1g-dev liblzma-dev
```

Next, download a source distribution of Python from the site `https://python.org`.

Make a note of the file name, which will look like the following: `Python-3.x.x.tar.xz`, where `x.x` is a pair of numbers that specify the build for this distribution. You should download the highest version available, which should be above 3.6.0.

Now, go to the directory where the downloaded file was saved and run the following command, replacing `x.x` with the corresponding build number of your file:

```
tar xvf Python-3.x.x.tar.xz
```

To build the distribution, execute the following commands, again replacing `x.x` with the correct build number:

```
cd Python-3.x.x
./configure --prefix=$HOME/python3
make
```

The build process will take a while to finish, depending on the configuration and speed of your system.

The following step is optional. If you want to run the battery of tests included with the source distribution, run the following command from the Terminal window:

```
make test
```

Depending on how fast your computer is, this may take a long time. At the end of the tests, a report of the build process will be generated. Do not worry if you get a few messages about skipped tests.

Next, install the code in its target directory, by running the following command:

```
make install
```

We now need to adjust the `PATH` environment variable. Start by making a backup copy of your shell configuration file by running the following commands:

```
cd
cp .bashrc .bashrc.python.bak
```

Start your text editor, open the `.bashrc` file, and add the following line at the end of the file:

```
export PATH=$HOME/python3/bin:$PATH
```

Save the file, close the editor and, back at the Terminal window, run the following command:

```
source ~/.bashrc
```

To test the installation, start Python from the command line by running the following command:

```
python3
```

If the installation is correct, Python will start and print information about the interpreter, as follows:

```
Python 3.x.x (default, Apr  4 2017, 09:40:21)
[GCC 4.2.1 Compatible Apple LLVM 8.1.0 (clang-802.0.38)] on darwin
Type "help", "copyright", "credits" or "license" for more information.
```

Check that the correct version of Python is being started by verifying that the version number 3.x.x coincides with that of the distribution you downloaded.

If you get an error when trying to start Python, or the wrong version of Python starts, the PATH variable was probably not set correctly in the `.bash_rc` file. You can check the current value of PATH by entering the echo $PATH command. If the path is not correct, edit `.bashrc` as indicated previously.

Exit Python by running the following command at the >>> Python prompt:

```
quit()
```

As a final test, run the following statement from the command line to check that Python's package manager was installed:

```
pip3 --version
```

This will print information about the version of pip3 that you are running. If no error message is issued, the setup was completed correctly and you can proceed to installing SciPy.

Installing the SciPy stack

Open a Terminal window and enter each of the following commands in succession:

```
pip3 install --user numpy scipy matplotlib
pip3 install --user ipython jupyter PyQt5
pip3 install --user pandas sympy nose
```

Make a backup copy of the .bashrc file and open it with a text editor. For example, to use nano, run the following commands in a Terminal window:

```
cd
cp .bashrc .bashrc.scipy.bak
nano .bashrc
```

Add the following line at the end of .bashrc:

```
export PATH="$HOME/.local/bin:$PATH"
```

Save the file, close the editor, and run the following command from the Terminal window:

```
source .bashrc
```

Let's now test the installation. Start Python by running the following command in a Terminal window:

```
python3
```

Execute the following lines at the >>> Python prompt. There will be no output and, as long as there are no errors, the installation is correct:

```
import numpy
import scipy
import matplotlib
import IPython
import pandas
import sympy
import nose
```

Exit Python by running the following at the prompt:

```
quit()
```

Back at the Terminal prompt, run the following command:

```
ipython
```

This will start IPython, an alternative Python shell with many added features, and print a startup message. You can check that the Python shell is running the expected version of Python from the top line of the startup message. Exit IPython by entering, at the prompt, the following command:

```
quit()
```

Back at the Command Prompt, run the following from the command line:

```
jupyter notebook
```

If all is correct, the Jupyter Notebook will start on your browser after a while. You have now concluded the installation of Python 3 and the SciPy stack in Linux.

Installing optional packages with conda

Sooner or later, you will need a package that is not installed by default in your Anaconda distribution.

If you are using the Anaconda distribution, this recipe describes the preferred method of installing and uninstalling packages. Anaconda is built on top of conda, a flexible and easy-to-use package manager.

Anaconda allows direct access to vast collections of add-on packages. For example, it makes it easy to install SciKits, which are SciPy packages developed by independent developers that are not part of the main distribution. A complete list of SciKits can be found at the following site: http://scikits.appspot.com/scikits.

Getting ready

This recipe assumes that you have a working installation of Anaconda. If you don't, follow the recipe for installing Anaconda for your operating system presented previously in this chapter.

How to do it...

As an example, lets install scikit-image, a package for image processing available at the site http://scikit-image.org:

To install `scikit-image`, first update `conda` and Anaconda by running the following two commands in your Terminal:

```
conda update conda
conda update anaconda
```

Now, `scikit-image` can be installed by running the following command at the system prompt:

```
conda install scikit-image
```

After the installation is finished, use a text editor to create a file called `skimage_test.py` containing the following code:

```
from skimage import data, io, filters

image = data.coins()
edges = filters.sobel(image)
io.imshow(edges)
io.show()
```

This code first constructs an image object from one of the sample images contained in `scikit-image`, which in this example is a photograph of a collection of old coins. It then applies a Sobel filter, which emphasizes the edges of objects present in the image. Finally, it displays the image of the detected coin edges.

Save the file and run the script by entering the following statement at the command line:

```
python3 skimage_test.py
```

Once the script runs, an image of the coin edges is displayed. Close the image to exit the script.

Let's now suppose that you don't need `scikit-image` any longer. It is easy to uninstall the package with `conda` using the following command:

```
conda uninstall scikit-image
```

If, after uninstalling, you attempt to run the `skimage_test.py` script, you will get an error message stating that Python can't import the `skimage` library.

Finally, what if a package is not available in Anaconda? In this case, it is still possible to install the package with `pip`, as outlined in the next section.

An alternative to installing packages directly in the global Anaconda distribution is to create a virtual environment with `conda`. Later in this chapter, we present several recipes for creating virtual environments with different requirements.

Installing packages with pip

The main tool to set up packages in a standalone installation of Python 3 is the `pip3` command. This command will fetch a package from the **Python Package Index** (**PyPI**), a standard public repository located at `https://pypi.python.org/pypi`.

PyPI is somewhat hard to navigate due to the large number of packages from different application areas that are available. A list restricted to packages related to SciPy is available at: `https://www.scipy.org/topical-software.html`.

The existence of a uniform package manager makes it is straightforward to add and remove packages to a Python installation.

 Python has a standard package distribution system, documented at the site `https://packaging.python.org`.
The procedures outlined in this site are the preferred methods to distribute Python software. This is recommended even in the case of software intended for internal use in an institution.

 If you are using Anaconda, the preferred method for installing packages is to use `conda`, as explained in the previous section. Only use `pip3` if the package is not available in Anaconda.

How to do it...

As an example, let's say we want to install the Bokeh package, located at `http://bokeh.pydata.org/`.

Bokeh is a visualization library that targets modern browsers for presentation. It produces high-quality plots in a variety of formats for inclusion in web pages. To install Bokeh in macOS or Linux, all you need to do is to run the following command in a Terminal window:

```
pip3 install bokeh
```

To install on Windows, run the following command at the Command Prompt:

```
pip install bokeh
```

This will remotely access PyPI, fetch Bokeh and all its dependencies, build the package, and make it accessible in the current Python 3 installation. To test the package, use a text editor to enter the following code in the `bokeh_test.py` file:

```python
import numpy as np
from bokeh.plotting import figure, show

xvalues = np.linspace(-np.pi, np.pi, 50)
y1values = np.sin(xvalues)
y2values = np.cos(xvalues)

p = figure(
 tools='pan,box_zoom,reset',
 title='Trigonometric functions',
 x_axis_label='x',
 y_axis_label='y'
 )

p.line(xvalues, y1values,
 legend='y=sin(x)', line_color='blue')
p.circle(xvalues, y2values,
 legend='y=cos(x)', fill_color='green',
 line_color='green', size=3)

show(p)

print('Plot generated. Open file bokeh_test.html in your browser.')
```

This code uses the `np.sin()` and `np.cos()` NumPy functions to generate data points for the trigonometric functions *sin(x)* and *cos(x)*. After that, it constructs a Bokeh `figure` object and adds two plots to the figure with the `p.line()` and `p.circle()` functions. Finally, the figure is displayed with a call to `show(p)`.

Save the file in a convenient folder and open a Terminal window from the same folder. Run the following statement at the command line:

```
python3 bokeh_test.py
```

Running the script will create an output file named `bokeh_test.html`, with an HTML code that generates the figure described in the Python script. If you have a default browser defined for your system, Bokeh will automatically open the HTML file in the browser. Otherwise, you will need to open the file in the browser manually.

Notice that the graph displays control buttons at the top right. By clicking on the **Pan** button, the user can drag the graphed region with the mouse. With the Box Zoom tool, it is possible to draw a box on the graph, and the plot will be zoomed in on that box. Finally, Reset brings the figure back to its original state.

Let's say that Bokeh does not satisfy your needs and you do not want to keep it in your Python installation. The package can be easily uninstalled by issuing the following command in the Terminal, for Linux and macOS:

```
pip3 uninstall bokeh
```

On Windows, use the following command:

```
pip uninstall bokeh
```

The uninstaller will list the changes to be made and ask for confirmation. If you answer yes, the package and its dependencies will be removed from the system.

> The `pip3` package manager is a wonderful and well designed tool. However, it is a complex tool, since it has to carefully keep track of the changes and dependencies in the Python installation. Furthermore, it depends on developers to structure distributed packages with the correct dependencies. If all the reader wants is to test a package, I recommend the use of a virtual environment, as explained in the next section.

Setting up a virtual environment with conda

Setting up a virtual environments with `conda` is very easy, and is recommended even for small projects. Virtual environments are very handy.

Getting ready

This recipe assumes that you have a working installation of Anaconda. If you don't, follow the recipe for installing Anaconda on your operating system presented previously in this chapter.

How to do it...

One of the features of `conda`, the standard package manager used with Anaconda, offers the easy creation and management of virtual environments. We will show you three recipes presenting the typical uses of `conda` virtual environments.

Before diving into the examples, enter the following statement in the command line:

```
conda info
```

This will print information about the current Anaconda installation. To obtain a list of the existing `conda` environments, enter the following:

```
conda info --envs
```

When I run this, I get the following output:

```
# conda environments:
#
root                   *   /Users/luizmartins/anaconda
```

Right now, there is only the root environment, since this is a fresh Anaconda installation.

Creating a virtual environment for development with conda

This recipe demonstrates how to set up an environment that is a clone of the current Anaconda installation. One possible use of this procedure is to set up an environment when we start the development of a new project.

Getting ready

This recipe assumes that you have a working installation of Anaconda. If you don't, follow the recipe for installing Anaconda on your operating system presented previously in this chapter.

How to do it...

1. Start by opening a Terminal window and running the following commands:

```
conda update conda
conda update anaconda
```

These commands update both `conda` and Anaconda to the most recent versions. This is always recommended before creating a new environment.

2. Next, run the following line to create a new environment named `myenv`:

```
conda create --name myenv
```

3. You will be asked for confirmation and `conda` will then create the new environment. We now need to activate the new environment. For Linux and macOS, run the following command in the terminal:

```
source activate myenv
```

4. In Windows, run the following:

```
activate myenv
```

5. Notice that the command line changes to reflect the active environment, as shown in the following example:

```
(myenv) computer:~ username
```

6. To confirm that the new environment was created, we can execute the following command:

```
conda info --envs
```

7. This command now produces the following output:

```
# conda environments:
#
myenv                 *   /Users/username/anaconda/envs/myenv
root                      /Users/username/anaconda
```

Notice that the currently active environment is marked with an asterisk.

8. We can now install packages in the active environment without interfering with the original Python distribution. For example, to install the `csvkit` package, we use code in the following example:

```
conda install csvkit
```

9. When you are done working in the environment, it is necessary to deactivate it. In Linux and macOS, this is done with the following command:

```
source deactivate
```

10. In Windows, the command to do so is the following:

```
deactivate
```

11. If you decide you don't need the `myenv` environment any longer, it can be deleted with the following command:

```
conda remove myenv --all
```

Note that you cannot remove the currently active environment.

Creating a conda environment with a different version of a package

We often need to run code that is not compatible with a particular version of a package. In this situation, it is useful to use `conda` to find out what version of a package is currently installed. For instance, to get information about which version of the `numpy` package is being used at the moment, we can execute the command shown as follows at the system prompt:

```
conda info numpy
```

On my computer, this outputs the following information:

```
numpy 1.12.1 py36_nomkl_0
file name   : numpy-1.12.1-py36_nomkl_0.tar.bz2
 name       : numpy
version     : 1.12.1
```

Getting ready

This recipe assumes that you have a working installation of Anaconda. If you don't, follow the recipe for installing Anaconda on your operating system presented previously in this chapter.

How to do it...

Let's now suppose that we have a legacy package that depends on an earlier version of NumPy. For example, let's assume we need version 1.7 of NumPy. We can check what versions of NumPy are available in the Anaconda repository with the following command:

```
conda search numpy
```

Running this command will produce a long list, in which we find the following information:

```
...
1.7.1                   py33_0  defaults
...
1.7.1                   py33_2  defaults
...
```

conda shows several available packages of NumPy, some of them compatible with Python 3. Let's now create an environment that has the older version of NumPy installed. Start by creating the new environment with the following command:

```
conda create --name numpy17 python=3 numpy=1.7
```

Notice that we specify the required versions for both Python and NumPy. conda will then display the following information about the environment to be created:

```
The following NEW packages will be INSTALLED:
numpy:    1.7.1-py33_2
openssl:  1.0.1k-1
pip:      8.0.3-py33_0
python:   3.3.5-3
readline: 6.2-2
  ...
```

Notice that conda will *downgrade* the version of Python in the new environment. conda packages are carefully designed to prevent incompatibilities. Go ahead and accept the changes to start the installation.

When the installation is complete, activate the new package. On Linux and macOS, we do this with with the following command:

```
source activate numpy17
```

On Windows, use the following command to activate the environment:

```
activate numpy17
```

After the environment is activated, we can install any packages that require version 1.7 of NumPy.

When we have finished working on the project, we need to deactivate the environment. On Linux and macOS, we use the following command:

```
source deactivate numpy17
```

On Windows, the command to deactivate an environment is as follows:

```
deactivate numpy17
```

Using conda environments to run different versions of Python

Another situation that comes up often is the need to test a package in a different version of Python. conda makes it easy to create a suitable environment. In this recipe, we will show you how to create and use environments with version 2.7 of Python.

Getting ready

This recipe assumes that you have a working installation of Anaconda. If you don't, follow the recipe for installing Anaconda on your operating system presented previously in this chapter.

How to do it...

1. Create the new environment with the following command:

   ```
   conda create --name python27 python=2.7
   ```

   conda will display information about the changes and ask for confirmation. Go ahead and enter yes to accept the environment creation.

2. To activate the environment on Linux and macOS, use the following command:

   ```
   source activate python27
   ```

3. On Windows, the command to activate an environment is as follows:

   ```
   activate python27
   ```

4. Let's now test the new environment. Start Python in the usual way after activating the environment. Notice that in the startup message, it should state that version 2.7 of Python is being run. Now run the following statement in the Python prompt:

   ```
   zip([['a','b'],[1,2]])
   ```

5. This will produce the following output:

   ```
   [('a', 1), ('b', 2)]
   ```

6. This confirms that we are running Python 2, since in Python 3 the zip() function returns an iterator object instead of a list. To leave the Python shell, run the following in the Python prompt:

   ```
   quit()
   ```

7. When done with the environment, deactivate it by running the following command on Linux and macOS:

   ```
   source deactivate python27
   ```

8. On Windows, run the following:

   ```
   deactivate python27
   ```

Creating virtual environments with venv

A *virtual environment* is a clone of a Python installation stored in a local folder. The Python executables and packages are not really copied, but symbolic links to the original files are created and environment variables are adjusted to set up a consistent Python filesystem. Once activated, a virtual environment will install packages in local directories without modifying the global Python installation.

A virtual environment is also the recommended setup for development in Python. With a virtual environment, it becomes easy to isolate the packages required to install the software under development.

 If you are using Anaconda, do not create virtual environments using venv. Instead, use conda, as described in the previous section.

How to do it...

1. To create the new environment in the myenv directory, use the following command in Linux and macOS:

    ```
    python3 -m venv myenv
    ```

2. In Windows, the command to be used is as follows:

    ```
    python -m venv myenv
    ```

    This creates an environment in the myenv subdirectory of the current directory. The -m switch is used because venv is a Python module that is being executed.

3. To switch to the new environment on Linux and macOS, run the following in the command line:

    ```
    source myenv/bin/activate
    ```

4. On Windows, we use the following command:

    ```
    myenv\bin\activate
    ```

5. If we need to install a package in the new environment, we use `pip` or `pip3` as usual. For example, to install a `molecule` package using `pip3`, we execute the following command:

   `pip3 install molecule`

6. Optionally, run the following command to create a `requirements` file:

   `pip3 freeze > requirements.txt`

   This will generate a report listing the packages required by the current environment and store the information in the `requirements.txt` file. This file can be distributed with the package being developed to facilitate installation by users that use `pip`.

7. When done working in the environment, deactivate it by running the following command:

   `deactivate`

8. If we no longer need an environment, we can delete it as follows. First, activate the environment using the following commands:
   On Linux and macOS, use the following:

   `source myenv/bin/activate`

   On Windows, use the following:

   `myenv\bin\activate`

9. Create a list of all packages in the environment with the following command:

   `pip3 freeze > requirements.txt`

10. Remove all packages installed in the environment with the following command:

    `pip3 uninstall -y -r requirements.txt`

11. The `-y` switch prevents `pip` from asking for confirmation to uninstall each individual package. Next, deactivate the environment by running the following command:

    `deactivate`

12. Finally, remove the environment directory with the following commands:

On Linux and macOS use the following:

```
rm -rf myenv
```

On Windows use the following:

```
rmdir /s /q myenv
```

It is important to uninstall all packages using `pip3` because some packages will copy files in directories outside of the environment directory. Typically, these directories are:
On Linux and Windows: The `.local` folder located in the user's home folder.
On macOS: `/Users/username/Library/Python/x.x/bin`, where `username` is the current user name and `x.x` denotes the version of Python the environment is based on.

Running SciPy in a script

Executing a program from a text file is the most time-honored approach to running computer code. Since Python is an interpreted language, text files meant to be run with Python are called *scripts*. Scripts are an easy way to share and distribute your programs, since all code is encapsulated in a number of files that can be easily copied to another user's computer.

Getting ready

To follow this recipe, you need a text editor and and a Terminal window. The Terminal window must be open on the directory where your script is saved.

How to do it...

Running a script in Python is only a matter of typing the code with a text editor and running it using the Python interpreter. Enter the following sample code into the text editor:

```
import numpy as np
from mpl_toolkits.mplot3d import Axes3D
import matplotlib.pyplot as plt
```

```
from matplotlib import cm

plt.switch_backend('Qt5Agg')

fig = plt.figure()
ax = fig.gca(projection='3d')
ax.view_init(elev=25, azim=65)
xvalues = np.linspace(-4, 4, 40)
yvalues = np.linspace(-4, 4, 40)
xgrid, ygrid = np.meshgrid(xvalues, yvalues)
zvalues = xgrid**3 * ygrid + xgrid * ygrid**3
ax.plot_surface(xgrid, ygrid, zvalues, cmap=cm.coolwarm,
                linewidth=0, antialiased=True)

plt.show()
```

Save the script in a file named `script_test.py`. Open a Terminal window on the directory where the script was saved and execute the following command from the system prompt:

python3 script_test.py

Running the script will produce a three-dimensional plot similar to the one displayed in the following screenshot:

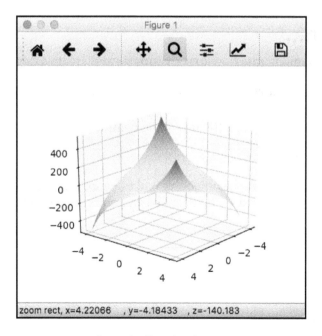

Image produced by running script_test.py

The image displayed is interactive, and can be rotated and panned with the mouse. It is also possible to save the image to the disk. Notice that the script is suspended while the image is being displayed. To continue execution of the script, simply close the image. In our example, this will cause the script to end.

Running SciPy in Jupyter

If your main goal in using SciPy is to do data exploration and analysis or scientific computations, Jupyter provides an ideal interactive environment. Using Jupyter, we can integrate computations, graphs, formatted text, and even more sophisticated media. Essentially, anything that can be inserted in a web page can be handled by Jupyter.

Getting ready

This recipe assumes that you have a working installation of IPython and Jupyter. If you followed one of the recipes in this chapter to set up Anaconda or a standalone installation of the SciPy stack, you have all you need.

How to do it...

The following steps demonstrate how to start Jupyter and create a new notebook:

1. Open a command window on the directory where you want your notebook files stored.
2. Start Jupyter by running the following command in the Terminal window:

   ```
   jupyter notebook
   ```

3. After a few moments, the notebook will open in your web browser. The notebook starting page is known as the dashboard, and is shown in the following screenshot:

4. To create a new notebook, click the **New** button at the top right and select **Python 3** from the menu. The following screenshot shows a newly created notebook:

Code in a notebook is entered in an *execution cell*, which is surrounded by a green border when active. To get a taste of what working with the Jupyter notebook feels like, click on an execution cell in the notebook and enter the following code:

```
%matplotlib inline
import numpy as np
import matplotlib.pyplot as plt
```

With the mouse cursor still in the same execution cell, press *Shift + Enter* to run the cell. The preceding code displayed the *magic* `%matplotlib inline` first to tell Jupyter that we want to display plots in the notebook itself, and the next two lines imported NumPy and pyplot (an interactive plotting library provided by `matplotlib`).

After running the cell, the cursor automatically moves to the next cell. Enter the following code in this cell:

```
from scipy.stats import norm, binom
n, p = 100, 0.5
mean = n * p
sdev = np.sqrt(n * p * (1-p))
sample = np.array([binom.rvs(n, p) for _ in range(1000)])
xvalues = np.linspace(mean-3*sdev, mean+3*sdev, 200)
yvalues = norm.pdf(xvalues, loc=mean, scale=sdev)
hist = plt.hist(sample, normed=True,
                color='red', lw=3, ls='dotted', alpha=0.5)
plt.plot(xvalues, yvalues, color='blue', lw=2)
plt.title('Coin toss simulation, $n={}$, $p={:5.2f}$'.format(n, p))
plt.xlabel('Number of heads')
plt.ylabel('Frequency')
None
```

This code simulates 100 tosses of a fair coin. The simulation is repeated 1,000 times and the results are stored in the array sample. Then a histogram of the results is plotted, together with a normal approximation, according to the central limit theorem. Pressing *Shift + Enter* to run the cell will produce a plot of the histogram, representing the simulation and the theoretical normal approximation of the distribution of the number of heads in the coin tosses.

Running SciPy in Spyder

Spyder—which is an acronym for **Scientific PYthon Development EnviRonment**—is an IDE specifically designed for Python and SciPy. It provides an out-of-the box solution for developing projects that use NumPy, SciPy, pandas, and other scientific or data-oriented Python libraries.

Getting ready

Spyder is included in the default Anaconda distribution, so, if you are using Anaconda, you already have Spyder.

If not, you can install it with `pip` using the following commands:

```
pip3 install PyQt5
pip3 install spyder
```

Note that depending on your configuration, you may already have PyQt5 installed, but attempting to install it again will do no harm. `pip3` will only tell you that the package is already present.

How to do it...

The first step is to start Spyder by running the following command from the command line:

```
spyder
```

After a few moments, the Spyder home window will appear on your screen, displaying an environment similar to the one shown in the following screenshot:

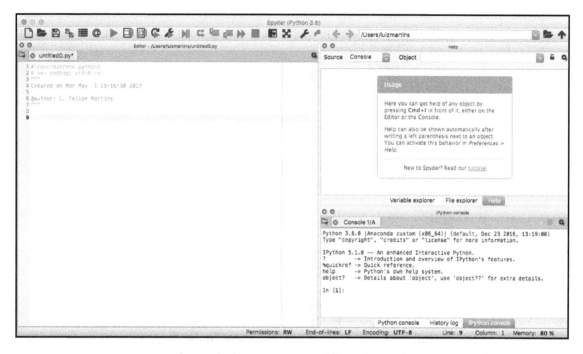

The window shown contains the following panes:

- On the left side of the window is the the **Editor** pane, where you enter code.
- The top right pane is the inspector pane. This pane is used to display information about code objects.
- The bottom right pane is the console pane. Notice that you can choose between an IPython console, a plain Python console, and the History Log.

To test the running code in Spyder, remove all text from the **Editor** window and type in the following code:

```python
import numpy as np
import matplotlib.pyplot as plt
from scipy.special import eval_chebyt

fig = plt.figure()
xvalues = np.linspace(-1, 1, 300)
nmax = 5
for n in range(nmax+1):
    yvalues = eval_chebyt(n, xvalues)
    plt.plot(xvalues, yvalues)
plt.title('Chebyshev polynomials $T_n(x)$ for
$n=0,\\ldots,{}$'.format(nmax))
plt.axhline(0, color='gray')
plt.axvline(0, color='gray')
plt.xlabel('$x$')
plt.ylabel('$T_n(x)$')

print('Displaying graph')
plt.show()
```

This code first imports NumPy, pyplot, and the special `eval_chebyt` function, which computes the values of Chebyshev polynomials of the first kind. Chebyshev polynomials are a family of orthogonal polynomials that are important in the design of some numerical methods. Then, the code defines a `figure` object, stored in the variable `fig`. Then it computes arrays containing values of the first six Chebyshev polynomials and plots them in the figure, which is then displayed after the formatting of the title and labels on the graph.

To run the code, go to the Spyder menu and select **Run-Configure**. The following window with configuration options will pop up on the screen:

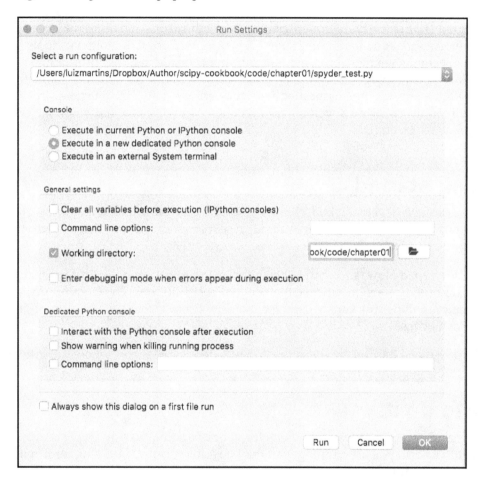

In the **Console** section, choose **Execute in a new dedicated Python console**, and leave all other options with their default values. Click **OK** to dismiss the box. Now, back in the main Spyder window, choose **Run-Run** on the menu, or press *F5*. The script will be launched and, after a few seconds, a window with a plot of the Chebyshev polynomials will be displayed. Closing the graph window will continue the execution of the script.

Spyder is a powerful yet simple to use IDE. To explore the features of Spyder, the reader is encouraged to explore the software's official documentation: https://pythonhosted.org/spyder/.

Running SciPy in PyCharm

For projects of moderate to large size, using a full-fledged **Integrated Development Environment (IDE)** is essential for efficient code production. An IDE might look, at first, like an imposing tool due to its complexity. A typical modern IDE integrates tools for editing, project management, building and running code, debugging, profiling, packaging for distribution, and integration with version control systems. As with any sophisticated tool, there is a learning curve associated with mastering a new IDE. The gains in productivity in the software development process, however, easily outweigh the time spent learning how to use an IDE.

Getting started

PyCharm is an IDE developed by JetBrains, providing a well-rounded collection of tools designed with Python development in mind. To experiment with PyCharm, go to the following address: `https://www.jetbrains.com/pycharm/`.

Click on **Download Now**, and you will be given a choice between the Professional Version and the Community Version. Choose the one that best suits you. The Community Version has limited capabilities but can be used for free. The Professional Version requires the purchase of a license.

After the download has finished, run the downloaded installer and follow the instructions.

How to do it...

Start PyCharm according to the directions for your operating system. Depending on the version of PyCharm you are running, you may be asked to enter licensing information. After that, the startup window will show up on your screen, as shown in the following screenshot:

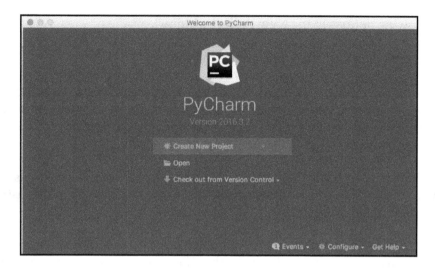

PyCharm requires the definition of a Project containing all files related to a particular development project. This may seem like overkill if we just want to run a single script, but PyCharm's project structure is very flexible. It does not require the setting up of a main entry point, and allows for several running configurations. This way, it is easy to run and debug individual scripts independently.

Click on **Create New Project** on the startup window, and the **New Project** window will pop up, as shown in the following screenshot:

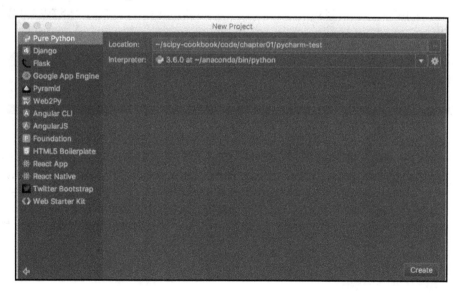

Select the following options in the **New Project** window:

In the **Project type** pane at the left, select **Pure Python**.

In the **Location** field, enter the directory where you want the project to be saved. You can use the button at the right, labeled with **...**, to choose the directory visually. If the directory does not exist, it will be created for you. In this example, we are using pycharm-test as the directory for our project. This will also be the name of the project in PyCharm.

Choose the Python interpreter for the project. For the purposes of this book, this should be the interpreter in the Python environment that was installed according to the instructions given previously in this chapter.

It is very important that you choose the correct environment for the project you are starting. Besides using the chosen interpreter for running code, some key features of PyCharm, such as Intellisense code completion, use this information to analyze your code. In particular, if you are using a virtual environment, make sure you select the Python interpreter corresponding to the correct environment.

Click the **Create** button, and the project creation process will begin.

If this is the first time you are running PyCharm, you may notice that, at the bottom right of the PyCharm window, there is an indicator stating that there are processes running in the background:

This indicator appears because PyCharm is indexing the Python packages currently in your environment. This index is used by several PyCharm components, including Intellisense, syntax highlighting, and syntax checker. The indexing process may take up to a few minutes, depending on how fast your computer is and how many packages are installed in the Python tree. You can continue to use PyCharm during the indexing process, but you may observe degraded responsiveness. Everything should go back to normal once the indexing process is finished.

Let's now take a brief tour of the main PyCharm window, shown in the following screenshot:

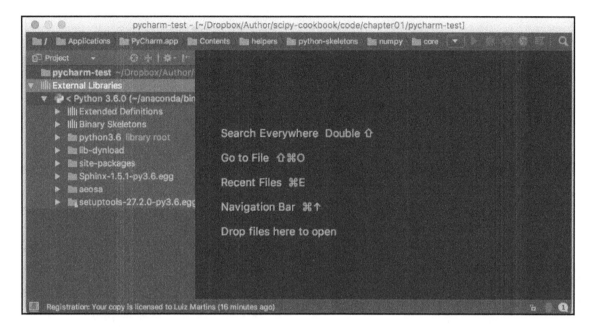

Right now, no files are being edited, and the right pane does not show any code. The left pane is currently showing the **Project Navigator**, from which all project components can be easily accessed. Right now, the project has no files.

Click on the arrow to the left of **External Libraries**. This gives you access to the Python tree corresponding to the Python interpreter that is being used in this project. Notice that the interpreter itself is listed right at the top of the list. You can use this pane to navigate through the Python installation, and open the source of any installed module in the editor .

Do not edit the source of any module in the Python distribution unless you know what you are doing. This information should be considered read-only in most cases. Changing the source of an installed module may break it, in which case you will probably need to scrap your whole Python tree and reinstall it.

Another useful component of the PyCharm window is the tool menu, represented by a little box on the extreme bottom left of the screen. Hover your mouse pointer over the box, and the menu shown in the following screenshot will be displayed:

This menu contains shortcuts to frequently used tools. If you work a lot with PyCharm, this menu is one of the best time savers offered by the IDE.

Let's now create some code and run it. Right-click on the project name, **pycharm-test**, on the **Project Navigator** and select **New** from the context menu. Then select **Python File** in the submenu that pops up. Enter the name of the file in the appropriate field, without the .py extension. Click **OK** and the file will be created and opened in the **Editor** pane.

Enter the following code in the pycharm-test file. To get a feel for PyCharm's coding features, it is recommended that you type the code yourself, instead of copying and pasting it to the **Editor** window:

```
def collatz_sequence(n):
    collatz_seq = [n]
    while n != 1:
        if n % 2 == 0:
            n //= 2
        else:
            n = 3 * n + 1
        collatz_seq.append(n)
    return collatz_seq


while True:
```

```
    n_input = input("Enter starting value (enter 0 to exit): ")
    try:
        n = int(n_input)
    except ValueError:
        print('Invalid integer: {}'.format(n_input))
        continue
    if n < 0:
        print('n must be positive')
    if n == 0:
        break
    cseq = collatz_sequence(n)
    print('Collatz sequence for n={}:\n{}'.format(n, cseq))

print('Thanks for playing, I hope you had fun.')
```

Notice the following as the code is typed:

- PyCharm will offer suggestions as you type the code. PyCharm analyzes your code on the fly, and searches the libraries and code you have already written for sensible suggestions as to what you are trying to type. To accept a suggestion, simply press *Enter*.
- If there are several alternatives, you can move up and down the list of suggestions with the arrow keys.
- Syntax errors are highlighted as they appear, and you can correct them before running the code.
- Suspicious code is also indicated. For example, if you define a variable or import and do not use it, PyCharm will warn you.

This code lets the user experiment with the famous Collatz problem, which is is an apparently simple problem that has so far resisted any approach anyone has tried. The problem consists of showing that a sequence constructed from a simple mathematical rule will always be finite, independently of the starting value of the sequence. Here, we simply use it as a example of code to be tested, but interested readers can read about the history of the problem at the following site: https://en.wikipedia.org/wiki/Collatz_conjecture.

To run the code, save the file and select **pycharm-test** on the **Project Navigator**. Right-click on **pycharm-test** and select **Run pycharm-test**. The script will start and, if there are no syntax errors, you will be prompted for a number on the PyCharm console, shown as follows:

```
Enter starting value (enter 0 to exit):
```

Enter a positive integer and press *Enter*. The script will compute the corresponding Collatz sequence and print it on the screen:

```
Enter starting value (enter 0 to exit): 7
Collatz sequence for n=7:
[7, 22, 11, 34, 17, 52, 26, 13, 40, 20, 10, 5, 16, 8, 4, 2, 1]
Enter starting value (enter 0 to exit): 22
Collatz sequence for n=22:
[22, 11, 34, 17, 52, 26, 13, 40, 20, 10, 5, 16, 8, 4, 2, 1]
```

After you are done experimenting with the script, enter 0 in the prompt line and the program will stop:

```
Enter starting value (enter 0 to exit): 0
Thanks for playing, I hope you had fun.
Process finished with exit code 0
```

Let's now use this code to experiment with the debugger. In the text editor, change the code that reads n = 3 * n + 1 to n = 2 * n + 1 and save the script. Run the code and enter a positive integer in the prompt. You will notice that no output is produced, and the script seems to be running in a loop. On the PyCharm menu, select **Run-Stop pycharm-test**. The script will be stopped and a KeyboardInterrupt exception will be reported in the console window. Stopping the script this way is equivalent to pressing *Ctrl + C* at the keyboard.

Let's set up a breakpoint to debug the code. We want to stop the script right after it enters the loop, at the line:

```
n_input = input("Enter starting value (enter 0 to exit): ")
```

To do this, click on the gutter at the left of the line, where we want to insert a breakpoint. The newly inserted breakpoint will be highlighted, as shown in the following screenshot:

```
def collatz_sequence(n):
    collatz_seq = [n]
    while n != 1:
        if n % 2 == 0:
            n //= 2
        else:
            n = 2 * n + 1
        collatz_seq.append(n)
    return collatz_seq


while True:
    n_input = input("Enter starting value (enter 0 to exit): ")
    try:
        n = int(n_input)
    except ValueError:
        print('Invalid integer: {}'.format(n_input))
        continue
    if n < 0:
        print('n must be positive')
    if n == 0:
        break
    cseq = collatz_sequence(n)
    print('Collatz sequence for n={}:\n{}'.format(n, cseq))
```

Let's now start the script in debugging mode, by right-clicking **pycharm-test** on the **Project Navigator** and selecting **Debug pycharm-test**. The script will start and break at the line that asks for user input. Select **Run-Step over** on the menu, or press the *F8* key, to run the next line in the script. When prompted, enter a positive integer and press *Enter* on the keyboard. At this point, the next line in the script will be highlighted:

```
try:
```

Continue to step over the code, until you get to the line:

```
cseq = collatz_sequence(n)
```

At this point, we want to go into the `collatz_sequence()` function to see what it is doing. To do this, select **Run-Step into**, or press *F7*. This will bring you into the code for `collatz_sequence()`. Now, continue to step over code (*F8*) until you can guess what is happening with the code.

You will notice that in the `if` statement inside the `while` loop, the `n = 2 * n + 1` statement is being repeatedly executed. Notice what happens in the **Editor** window. After a few runs through the loop, the editor will show something like the following screenshot:

```
pycharm-test.py
1
2   def collatz_sequence(n):   n: 1023
3       collatz_seq = [n]   collatz_seq: <class 'list'>: [7, 15, 31, 63, 127, 255, 511, 1023]
4       while n != 1:
5           if n % 2 == 0:
6               n //= 2
7           else:
8               n = 2 * n + 1
9           collatz_seq.append(n)
10      return collatz_seq
11
12
```

Notice that as you step over code, PyCharm displays an update about the contents of the `collate_seq` list. You will notice that this list keeps growing and, if you let the code run long enough, you will eventually get an exception due to a memory fault.

So, what is happening here? When the code gets to the `if` statement, it does the `n % 2 == 0` test. This returns `True` only when n is even. If n is odd, the code falls through the `else` clause, where n is updated by the `n = 2 * n + 1` statement. As a result, the updated value of n will again be an odd number and, the next time the loop body is executed, the code will again fall into the `else` clause. It is clear that this leads to an infinite loop.

Change the `n = 2 * n + 1` line back to `n = 3 * n + 1` to fix the code, and run it again to make sure everything is back as it should be.

The attentive reader will notice that, with the correct line of code, `n = 3 * n + 1`, if a particular value of n is odd, the next value of n will be even. Then, there will be a sequence of steps in which the number n is halved in the `n //= 2` statement, until we get another odd number. One of the difficulties with the Collatz problem is that it seems that the number of times that n is halved is quite unpredictable, even though the code is completely deterministic. The reader might want to investigate the problem by him or herself, but be warned, this can be quite a rabbit hole!

In the preceding examples, we have barely scratched the surface of all features offered by PyCharm. The reader is invited to try using PyCharm on one of his/her projects, and to read the documentation to learn all that is available.

2
Getting Started with NumPy

In this chapter, we will learn the following recipes:

- Creating NumPy arrays
- Querying and changing the shape of an array
- Storing and retrieving NumPy arrays
- Indexing
- Operations on arrays
- Using marked arrays to represent invalid data
- Using object arrays to store heterogeneous data
- Defining, symbolically, a function operating on arrays

Introduction

We are now ready to start exploring NumPy, the fundamental package upon which the whole scientific Python stack is built. This chapter presents an introduction to the essential features of NumPy that are used in day-to-day scientific and data computations.

Built-in Python data structures, such as lists and dictionaries, are ill suited for scientific and data-oriented computing and their use results in programs that are significantly slower than numerical code written in compiled languages such as C, C++, and Fortran. NumPy was created to address this problem, and solves it by defining specialized array-oriented objects and methods designed for efficient numerical and data computing. The main data structure defined in NumPy with this purpose is `ndarray`, which represents a multidimensional array of data.

Objects of `ndarray` type differ from the native Python data structures in the following respects:

- In general, data held in an `ndarray` must be of a uniform type, determined at the time the array is constructed.
- Arrays can hold a wider range of data types than Python. In general, all types supported by standard implementations of the C programming language are supported.
- Objects of `ndarray` type have fixed dimensions. Resizing or reshaping an array results in the creation of a copy.
- Objects of `ndarray` type have a rich set of methods for efficient manipulation of arrays, including creation, element reference and assignment, access to ranges, reshaping, and concatenation.
- A special kind of function, represented by the `ufunc` class, is defined to operate on arrays. A `ufunc` can apply the same function to all elements of an array of arbitrary shape..

Examples in this chapter were tested on release 1.12.0 of NumPy, which is the latest available version as of writing. The package documentation can be accessed at the site `https://docs.scipy.org/doc/`.

To run the examples in this chapter, it is recommended that you enter the code in an interactive Python environment. The options are as follows, in order of preference:

- The Jupyter Notebook, which can be started by running the command `jupyter notebook` at the terminal.
- The IPython shell, which can be started by running `ipython` at the command line.
- The Python shell, which can be started by running `python3` or `python` at the command line. Make sure you are running the shell corresponding to the installation of Python for which the scientific Python stack was installed.

In this chapter, we will present recipes for performing the following tasks:

- How to create NumPy arrays
- How to change the size and shape of an array
- How to concatenate, or *stack* arrays
- How to select elements or ranges of an array with different indexing mechanisms

- How to store/retrieve arrays to/from permanent storage
- How to apply vectorized functions and operations to arrays
- How to define vectorized functions using symbolic expressions

To run the examples in this chapter, it is first necessary to import the `numpy` module using the following code:

```
import numpy as np
```

Never use `from numpy import *` to import the `numpy` module. This would have the effect of importing hundreds of names into the currently running module, which could lead to hard to detect bugs due to name collisions.

Creating NumPy arrays

There are several ways to create objects of `ndarray` type. The recipes in this chapter provide a comprehensive list of the possibilities.

How to do it...

Let's move on to learn how an array can be created from a list.

Creating an array from a list

To create an array from an explicit list, use the following code:

```
x = np.array([2, 3.5, 5.2, 7.3])
```

This will assign to x the following array object:

```
array([ 2. , 3.5, -1. , 7.3, 0. ])
```

Notice that integer array entries are converted to floating point values. NumPy arrays are *homogeneous*, that is, all elements of an array must have the same type. Upon creation, elements in the input list are converted to a common type by a process known as *casting*. In the preceding example, all elements are cast to floats.

To create a multidimensional array, use a list of lists:

```
A = np.array([[1, -3, 2],[2, 0, 1]])
```

This creates the array:

```
array([[ 1, -3, 2],
       [ 2, 0, 1]])
```

The array elements in this example are integers. Creating arrays with more than two dimensions using this method is unwieldy, and is not recommended. Instead of a list literal, a list comprehension can be used for initialization:

```
x = np.array([i**2 for i in range(5)])
```

This creates an array containing the squares of the integers in the range of 0 to 4:

```
array([ 0,  1,  4,  9, 16])
```

This is not the most efficient way to generate this array in NumPy. We will later see how to use vectorized functions to generate this array very efficiently.

Specifying the data type for elements in an array

To create an array with elements of a given type, specify the dtype option on the array constructor, as follows:

```
x = np.array([1.2, 3.4, -2.0], dtype=np.float32)
```

This creates an array of 32-bit floating point values:

```
array([ 1.20000005,  3.4000001 , -2.        ], dtype=float32)
```

Types of array elements are represented in NumPy by objects from the dtype class, which stands for data type. The following table lists the most commonly used numeric data types available in NumPy:

Object	Data type
bool_	Boolean, stored as a byte (0 is FALSE, 1 is TRUE)
int32	32-bit signed integer
int64	64-bit signed integer
uint32	32-bit unsigned integer

`uint64`	64-bit unsigned integer
`float32`	32-bit floating point number (single precision float)
`float64`	64-bit floating point number (double precision float)
`complex64`	Complex number, represented by two 32-bit floats
`complex128`	Complex number, represented by two 64-bit floats
`object`	Arbitrary Python objects

If a `dtype` is not explicitly given in the constructor, NumPy will choose the native Python type that can best represent the data. Thus, in a 64-bit architecture, Python floats are converted to `float64` and Python integers are usually converted to `int64`.

Python integers are arbitrary precision integers, that is, they can have an arbitrary number of digits. If a Python integer is too large to be represented as a NumPy integer type, a NumPy `object` array will be created by default. This is usually not what we want in numerical computation, and in this case it is recommended that you use an explicit type. For example, to create an array with the single 2^{300} element, represented as a floating point number, use the code:
`x = np.array([2**300], dtype=np.float64)`.

Creating an empty array with a given shape

To create a 3 x 2 array of doubles, use the following code:

```
x = np.empty((3,2))
```

This creates the following array:

```
array([[ 0.,   0.],
       [ 0.,   0.],
       [ 0.,   0.]])
```

Notice that the array is not actually empty, but filled with arbitrary values. The `dtype` of the array is, by default, `float64`. A different type can be specified, as in the following example:

```
x = np.empty((2,4), dtype=np.int16)
```

This produces the following array:

```
array([[0, 0, 0, 0],
       [1, 0, 0, 0]], dtype=int16)
```

 It is important to notice that `empty()` will *not* initialize the array elements. The values will just be whatever is present in memory when the array is created. If you want an initialized array, use a function such as `zeros()`.

Creating arrays of zeros and ones with a single value

Let's see how to do it all with a single value.

An alternative to `empty()` is the `zeros()` function, which creates an array where all elements are initialized to zero, as in the following example:

```
x = np.zeros((2,3))
```

The `(2,3)` tuple argument states that we want a 2 x 3 array of zeros, so we get the following result:

```
array([[ 0.,   0.,   0.],
       [ 0.,   0.,   0.]])
```

Notice that, by default, the data type of the array is `float64`. If an alternate data type is desired, it must be specified in the call to `zeros()`, as indicated in the following code:

```
x = np.zeros((2,3), dtype=np.int64)
```

This will create a 2 x 3 array of integers.

The `ones()` function creates an array where all elements are initialized to one, shown as follows:

```
x = np.ones((4,4))
```

This command creates a 4 x 4 array of elements of `float64` type, all equal to 1. As was the case with the `zeros()` function, the data type can be specified by setting the `dtype` argument in the function call.

To create an array where all elements are an arbitrary value, use the `full()` function, shown as follows:

```
x = np.full((3,1), -2.5)
```

This creates a 3 x 1 array with all elements initialized to the value -2.5. The data type of the array will be `float64`.

Creating arrays with equally spaced values

There are two NumPy functions that create arrays with equally spaced elements: `arange()` and `linspace()`. The `arange()` function, which stands for *array range*, has an interface similar to the built-in Python `range()` function, except that the arguments can be floats, as shown in the following example:

```
x = np.arange(1.1, 1.9, 0.2)
```

This generates an array with elements between 1.1 and 1.9, with increments of 0.2, resulting in the array as follows:

```
array([ 1.1,  1.3,  1.5,  1.7])
```

Notice that the right endpoint is not included, that is, `arange()` uses the same convention as `range()`. As in the built-in Python `range()` function, the increment can be omitted, and is then assumed to be equal to 1, as demonstrated in the following code:

```
x = np.arange(2.5, 6.5)
```

Here, only the start and end point of the range is given, producing the array:

```
array([ 2.5,  3.5,  4.5,  5.5])
```

If only one argument is given, the initial point is assumed to be zero and the increment is assumed to be one, as shown in the following code:

```
x = np.arange(3.2)
```

This statement will generate the following array:

```
array([ 0.,  1.,  2.,  3.])
```

All the preceding examples generate ranges with the `float64` element type. To generate a range with a different type, we can either use integer arguments in the function call, or specify the `dtype` option, as shown in the following example:

```
x = np.arange(0.9, 4.9, dtype=np.int64)
```

This statement, somewhat surprisingly, generates the following array:

```
array([0, 1, 2, 3])
```

To understand the output, notice that when generating the array range, NumPy will first cast the functions argument to the int type, which always rounds towards 0. So, the preceding range is effectively identical to the following:

```
x = np.arange(0, 4)
```

An alternative to arange() is the linspace() function, which is illustrated in the following example:

```
x = np.linspace(1.2, 3.2, 11)
```

This statement generates an array of 11 equally spaced floats from 1.2 to 3.2, *including both endpoints*, producing the array as follows:

```
array([ 1.2,  1.4,  1.6,  1.8,  2. ,  2.2,  2.4,  2.6,  2.8,  3. ,  3.2])
```

Creating an array by repeating elements

To create arrays with repeated elements, we use the repeat() function, as shown in the following example:

```
np.repeat(1.2, 3)
```

This repeats the 1.2 value three times, resulting in the following array:

```
array([ 1.2,  1.2,  1.2])
```

The argument to be repeated can be itself an array, as shown in the following example:

```
x = np.array([[1,2,3],[4,5,6]])
y = np.repeat(x, 2)
```

This will store, in the y variable, the following output array:

```
array([1, 1, 2, 2, 3, 3, 4, 4, 5, 5, 6, 6])
```

Notice that this version of repeat() *flattens* the array, that is, it returns a one-dimensional ndarray. If we want to prevent flattening, we can use the axis option, demonstrated as follows:

```
y = np.repeat(x, 2, axis=0)
```

This produces the array:

```
array([[1, 2, 3],
       [1, 2, 3],
       [4, 5, 6],
       [4, 5, 6]])
```

To understand this output, it is important to understand the notion of `axis` in NumPy. This notion applies to general *n*-dimensional arrays, but, to keep the examples simple, let's consider the case of a 3 x 2 two-dimensional array. The array can be visualized as a table, as follows:

	Column 0	Column 1	Column 3
Row 0	x[0,0]	x[0,1]	x[0,2]
Row 1	x[1,0]	x[1,1]	x[1,2]

Each element in the array is indexed by a pair of integers, `[i,j]`. We say that `i` is the index along axis 0, and `j` is the index along axis 1. Pictorially, axis 0 goes vertically in the downward direction, across the rows, and axis 1 goes horizontally to the right, across the columns.

If we move along axis 0, the array consists of two row vectors, as follows:

Row 0	x[0,0]	x[0,1]	x[0,2]
Row 1	x[0,0]	x[0,1]	x[0,2]

Accordingly, the `np.repeat(x, 2, axis=0)` function call takes each of the rows of the `x` array, copies it twice, and stacks all resulting rows vertically. Similarly, we could have called `repeat` with `axis=1`, as in the following example:

```
y = np.repeat(x, 2, axis=1)
```

In this example, the `x` array is interpreted as a sequence of column vectors, and each column is repeated twice, producing the following result:

```
array([[1, 1, 2, 2, 3, 3],
       [4, 4, 5, 5, 6, 6]])
```

One final option in `repeat()` is to use a list to specify the number of repetitions of each row/column, as in the following example:

```
y = np.repeat(x, [2, 1, 2], axis=1)
```

Since we are specifying `axis=1`, the array x is viewed as a sequence of columns. Then, column 0 is repeated twice, column 1 is repeated once, and column 2 is repeated twice, resulting the following array:

```
array([[1, 1, 2, 3, 3],
       [4, 4, 5, 6, 6]])
```

Notice that by setting the number of repetitions to 0, it is possible to use `repeat()` to delete the rows/columns of an array, as shown in the following example:

```
x = np.array([[1,2,3,4],
              [5,6,7,8]])
y = np.repeat(x, [1,0,1,0], axis=1)
```

Since the number of repetitions for columns 1 and 3 is equal to zero, these columns will be removed from the array, resulting in the following output:

```
array([[1, 3],
       [5, 7]])
```

Creating an array by tiling another array

An alternative to `repeat()` is the `tile()` function, which is used to copy a whole array into a tiled pattern. For example, let's try the following:

```
x = np.array([[1,2],[3,4]])
y = np.tile(x,(2,3))
```

This will take the array x and repeat it twice along axis 0, and then repeat the resulting array three times along axis 1. The result is a 2 x 3 *tiling* of the input array, as shown in the following output:

```
array([[1, 2, 1, 2, 1, 2],
       [3, 4, 3, 4, 3, 4],
       [1, 2, 1, 2, 1, 2],
       [3, 4, 3, 4, 3, 4]])
```

Creating an array with the same shape as another array

NumPy provides a family of functions that create an array with the same shape as another input array. These functions all end with the _like suffix, and are listed in the following table:

Function	Description
empty_like()	Creates an empty array with same shape as a given array
zeros_like()	Creates an array of zeros with the same shape as a given array
ones_like()	Creates an array of ones with the same shape as a given array
full_like()	Creates an array of a repeated value with the same shape as a given array

For example, the following code creates an array of zeros with the same shape as the x input array:

```
x = np.array([[1,2],[3,4],[5,6]], dtype=np.float64)
y = np.zeros_like(x)
```

This code first defines a 3 x 2 array, x. Then, it creates an array of zeros with the same shape as x, resulting in a 3 x 2 array of zeros.

It is important to notice that the type of the array constructed by the _like function is the same as the input array. So, the following code produces a slightly unexpected result:

```
x = np.array([[1,1],[2,2]])
z = np.full_like(x, 6.2)
```

The preceding code creates the following array:

```
array([[6, 6],
       [6, 6]])
```

The programmer writing the previous code probably wanted to generate an array with the value 6.2 repeated in all positions. However, the array x in this example has a dtype of int64, so the full_like function will coerce the result to an array of 64-bit integers. To obtain the desired results, we can specify the type of array we want as follows:

```
z = np.full_like(x, 6.2, dtype=np.float64)
```

When in doubt, it is advisable to specify the data type of an array explicitly using the `dtype` option. As a matter of fact, I prefer to always specify the `dtype` of an array, to make my intentions clear.

Using object arrays to store heterogeneous data

Up to this point, we only considered arrays that contained native data types, such as floats or integers. If we need an array containing heterogeneous data, we can create an array with arbitrary Python objects as elements, as shown in the following code:

```
x = np.array([2.5, 'a string', [2,4], {'a':0, 'b':1}])
```

This will result in an array with the `np.object` data type, as indicated in the output line reproduced as follows:

```
array([2.5, 'string', [2, 4], {'a': 0, 'b': 1}], dtype=object)
```

We mentioned that all elements in a NumPy array must be of the same type. In the case of arrays of objects, NumPy wraps the data in each array item with a common object type. The objects are unwrapped when accessed, so that the conversion is transparent for the user.

If the objects to be contained in the array are not known at construction time, we can create an empty array of objects with the following code:

```
x = np.empty((2,2), dtype=np.object)
```

The first argument, `(2,2)`, in the call to `empty()`, specifies the shape of the array, and `dtype=np.object` says that we want an array of objects. The resulting array is not really empty but has every entry set as equal to `None`. We can then assign arbitrary objects to the entries of x.

In a NumPy array of objects, as in Python lists and tuples, the stored values are *references* to the objects, not copies of the objects themselves.

See also

Joining arrays

Another way to create new arrays is to *join*, or *concatenate*, other arrays. There are several functions that make that easy. One common task is to create a two-dimensional array from its columns, and NumPy provides the `column_stack()` function for this specific purpose, as shown in the following example:

```
x = np.array([1,2,3])
y = np.array([4,5,6])
z = np.array([7,8,9])
w = np.column_stack([x,y,z])
```

This produces the array:

```
array([[1, 4, 7],
       [2, 5, 8],
       [3, 6, 9]])
```

 Notice that the arrays x, y, and z are interpreted as *column* vectors. The `columns_stack()` function always interprets its arguments as column arrays.

The `concatenate()` function is used to stack arrays along an arbitrary axis. The following example shows the concatenation of two arrays along axis 0:

```
x = np.array([[1,2,3],[4,5,6]])
y = np.array([[10,20,30],[40,50,60],[70,80,90]])
z = np.concatenate([x,y])
```

With this code, the arrays x and y are stacked vertically, generating the following array:

```
array([[ 1,  2,  3],
       [ 4,  5,  6],
       [10, 20, 30],
       [40, 50, 60],
       [70, 80, 90]])
```

Notice that, when using `concatenate()`, all input arrays must have the same dimension, except possibly for the axis along which they are being stacked. The following code shows how to concatenate two arrays side by side:

```
x = np.array([[1, 2, 3, 4],[5, 6, 7, 8]])
y = np.array([[10,20],[30,40]])
z = np.concatenate([x,y], axis=1)
```

The output will now be as follows:

```
array([[ 1,  2,  3,  4, 10, 20],
       [ 5,  6,  7,  8, 30, 40]])
```

The preceding examples of `concatenate()` represent the usual cases, that is, concatenating two-dimensional arrays along either axis 0 or axis 1. In these specific cases, NumPy has the convenient `vstack()` and `hstack()` functions. Thus, an alternative to stacking two arrays vertically is indicated in the following code:

```
x = np.array([[1,2,3],[4,5,6]])
y = np.array([[10,20,30],[40,50,60],[70,80,90]])
z = np.vstack([x,y])
```

Analogously, we can stack two arrays horizontally with the code:

```
x = np.array([[1, 2, 3, 4],[5, 6, 7, 8]])
y = np.array([[10,20],[30,40]])
z = np.hstack([x,y])
```

If we want to add elements at the end of an array, we can use the `append()` function. The following code example shows how to add a row at the bottom of an array:

```
x = np.array([[1,2,3],[4,5,6]])
y = np.array([[7,8,9]])
z = np.append(x, y , axis=0)
```

Notice that in the call to `append()`, we specifically state the concatenation axis with the `axis=0` option. If the axis is not specified, `append()` will flatten the output array, which is not normally what is wanted. Also, notice that the `y` array is defined with the `np.array([[7,8,9]])` expression, which is a 1 x 3 array. The code would not work if `y` were defined with `np.array([7,8,9])`, which does not match the shape of array `x`.

Querying and changing the shape of an array

Since we have learnt various ways to create an array, we can now definitely learn how to query and change the shape of an array.

How to do it...

The shape of an array is stored in the `shape` field of the `ndarray` object, as shown in the following example:

```
x = np.array([[1,2,3,4,5,6],[7,8,9,10,11,12]])
x.shape
```

The `shape` field of an `ndarray` object contains a tuple with the size of each of the dimensions of the array, so the preceding code will produce the following output:

```
(2, 6)
```

It is possible to assign a to the field shape, which has the effect of reshaping the array, as shown in the following example:

```
x.shape = (4,3)
```

This statement will change the array to the following:

```
array([[ 1,  2,  3],
       [ 4,  5,  6],
       [ 7,  8,  9],
       [10, 11, 12]])
```

Storing and retrieving NumPy arrays

Most realistic applications deal with large datasets and will require results to be stored in persistent media, such as a hard disk. NumPy arrays can be stored in either text or binary format and binary files can be optionally compressed. We start the section by showing you a recipe to store files in text format.

How to do it...

Let us proceed with getting our queries about storing and retrieving NumPy arrays resolved.

Storing a NumPy array in text format

To store a single NumPy array to the disk in text format, use the `savetxt()` function. In the next example, we generate a large array and save it to the disk in text format using the following code:

```
x = np.random.rand(200, 300)
np.savetxt('array_x.txt', x)
```

This code first uses the `np.random.rand()` function to generate a 200 x 300 array of random floats. Next, the `savetxt()` NumPy function creates the `array_x.txt` file on the disk, containing a text representation of the `x` array. The file can be opened with a text editor.

Saving NumPy arrays in text format always implies loss of precision, since the binary data has to be converted to decimal. Text formats also tend to require more disk space. Unless there is need to open the file with a different software package or the file needs to be human-readable, it is recommended that NumPy arrays are saved in binary format.

Storing a NumPy array in CSV format

The `savetxt()` function admits several parameters, which are useful when we want to output the array in a format that is compatible with a specific application. In this recipe, we show you how to store a NumPy array in CSV format. **Comma-separated values (CSV)** are used to store tabular data in a text file. Each row of the table is stored in a line of text and elements in each row are separated by a comma. The following code illustrates how to save an array in CSV format:

```
m, n = 10, 5
x = np.random.rand(m, n)
columns = ','.join(['Column {}'.format(str(i+1)) for i in range(n)])
np.savetxt('array_x.csv', x, fmt='%10.8f',
           delimiter=',', header=columns, comments='')
```

With this code, we first generate a 10 x 5 array x with random data. Next, we define a `columns` string containing the table headings, separated by commas. Finally, we save the array to the `array_x.csv` file. The options used in the call to `savetxt()` are described as follows:

- `fmt='%10.8f'` specifies that each array element should be formatted as a floating-point value in a field with a width of 10 and a precision of 8 decimals.
- `delimiter=','` sets the field separator to be a comma.
- `header=columns` directs `save_text()` to output the `columns` string at the top of the file. This effectively sets the headings for each column in the CSV file.
- `comments=''` is necessary because `save_text()`, by default, prepends a # to the headers line. This options tells us that an empty string should be prepended instead.

The output file, `array_x.csv`, can be opened with any spreadsheet software.

> The CSV file format is not standardized and each vendor uses slightly different conventions. More robust tools to create CSV files, as well as other data formats, are described in `Chapter 4`, *Data Wrangling with pandas*.

Loading an array from a text file

The `loadtxt()` function reads an array stored using the text format, in particular arrays stored using the `savetxt()` function. The following line of code reads the array stored in the `array_x.txt`, which was generated in a previous example:

```
x = np.loadtxt('array_x.txt')
```

This code will not produce any output, since the array is silently assigned to the x variable. To see a snapshot of the array, we can execute the following in the command line or Jupyter cell:

```
x[0:3, 0:3]
```

This prints the first three rows and columns of the array.

The `loadtxt()` command offers some flexibility in terms of the format of the text file being read, and has several options to facilitate loading data generated by other software. As an example, let's suppose that we have a text file with data from the following table:

Name	ID	Position	Salary	Years of service
Rob Reliable	101	Associate	42000.00	5
Sam Social	203	Associate	31000.00	3
Hellen Hardworking	105	Manager	67000.00	8

The data is in the `employees.txt` file and each row in the table occupies one line in the file, including the headings. In each row, fields are separated by a comma. Notice that some of the columns contain string data. Let's assume that we are only interested in the numerical fields. Under these assumptions, the relevant columns can be read into a NumPy array with the following statement:

```
x = np.loadtxt('employees.txt', delimiter=',',
               skiprows=1, usecols=(1,3,4))
```

The options in the `loadtxt()` function have the following meanings:

- `delimiter=','` specifies a comma as the field separator.
- `skiprows=1` states that the first row, containing the column headers, should be skipped.
- `usecols=(1,3,4)` specifies which columns of the table should be read into the array. Since indexing is zero-based, this would result in the second, fourth, and fifth columns being loaded.

Notice that, since NumPy arrays are restricted to holding elements of the same data type, we have to use an array of floats in this example, even though the ID and Years of service columns would be better represented by integers. We could use a NumPy array with `dtype=object` to store all columns, but a better approach would be to use a pandas `DataFrame`, which is a data structure designed with generic data in mind. pandas is looked at in detail in `Chapter 4`, *Data Wrangling with pandas*.

Storing a single array in binary format

NumPy defines a binary format called NPY to store arrays. Users do not need to know the details of the format, but those interested can find its specification at the following site: `https://docs.scipy.org/doc/numpy/neps/npy-format.html`.

The NPY format is designed with the efficient storage of NumPy arrays in mind, including optimizations for large arrays. It is a format specific to NumPy, so it is not a suitable exchange format for a web application, for example. It is designed for personal use, or for sharing data among coworkers in the same project. It does not offer any built-in security features and should not be used to share sensitive data. NumPy also defines the NPZ file format, which is simply a ZIP file packaging several NPY files.

To store a single array in binary NPY format, we use the `save()` function as follows:

```
x = np.random.rand(200, 300)
np.save('array_x.npy', x)
```

The preceding code will generate an array x with random data, and then call `save()` to create the `array_x.npy` file on the disk, containing a binary representation of the array x.

Storing several arrays in binary format

The `savez()` function allows for the saving of several arrays in the same file. In the following code, we generate arrays with random data and store them to the disk:

```
x = np.random.rand(200, 300)
y = np.random.rand(30)
z = np.random.rand(10, 5, 7)
np.savez('arrays_xyz.npz', xfile=x, yfile=y, zfile=z)
```

This code performs the following steps:

- Generates files in NPY format, holding each of the arrays x, y, and z. These files are named, respectively, `xfile`, `yfile`, and `zfile`.
- Creates a single ZIP archive containing the NPY files generated for each array.
- Saves the archive to the disk in a file named `arrays_xyz.npz`.

Since the generated archive is a ZIP file, it can be opened with any standard archiving utility. In my system, a printout of information on the contents of the `arrays_xyz.npz` file produces the following:

```
Archive:  arrays_xyz.npz
Zip file size: 483584 bytes, number of entries: 3
-rw-------  2.0 unx    480080 b- stor 17-May-09 09:32 xfile.npy
-rw-------  2.0 unx       320 b- stor 17-May-09 09:32 yfile.npy
-rw-------  2.0 unx      2880 b- stor 17-May-09 09:32 zfile.npy
3 files, 483280 bytes uncompressed, 483280 bytes compressed:  0.0%
```

This output indicates that the `arrays_xyz.npz` archive contains three files, `xfile.npy`, `yfile.npy`, and `zfile.npy`. Notice the last line of the report, stating that the compression factor is 0%. By default, the `storez()` function does not compress the data, which makes it work faster at the cost of producing larger files. If a compressed file is required, we can use the `savez_compressed()` function, demonstrated in the following code:

```
np.savez_compressed('arrays_xyz_c.npz', xfile=x, yfile=y, zfile=z)
```

By checking the generated file, `arrays_xyz_c.npz`, with an archive utility, it can be seen that the compression factor in this example is about 6%. The low compression rate is due to the fact that the arrays contain random data, which has little redundancy. Data exhibiting more regularity will yield better compression rates.

Loading arrays stored in NPY binary format

Independent of which function was used to generate a file containing NumPy arrays in NPY format, the `load()` function is used to retrieve the data from disk. In the following examples, we assume that the files `array_x.npy`, `arrays_xyz.npz`, and `arrays_xyz_c.npz` were generated, as in the previous examples.

The single array contained in the `array_x.npy` file can be loaded with the following command:

```
x = np.load('array_x.npy')
```

This will open the file, read the array into the variable x, and close the file.

Loading arrays saved with `savez()` and `savez_compressed()` is slightly more complicated. The following code shows how to load the arrays stored in the `arrays_xyz.npz` file:

```
arrays = np.load('arrays_xyz.npz')
x, y, z = [arrays[name] for name in ['xfile', 'yfile', 'zfile']]
```

For an `npz` file, the `load()` function returns a dictionary-like object. The keys to the dictionary are the file names in the archive, and the values are in the corresponding NumPy array. If the key names are not known, they can be recovered with the `keys()` method, shown as follows:

```
arrays.keys()
```

In our example, this returns the following:

```
['xfile', 'yfile', 'zfile']
```

Indexing

In this section, we address the methods NumPy offers for access and modification of data in an array. Python itself provides a rich set of indexing modes, and NumPy extends these with a number of methods suitable for numerical computations.

To access items of an array a, NumPy, as Python, uses the a[...] bracket notation. In the background, NumPy defines the __getitem__, __setitem__, and __deleteitem__ methods to do the requested operations on the array items. The arguments inside the brackets are expressions that specify the locations of the items we want to access. For example, to access the element at position (1,2) of the two-dimensional array a, we use the expression a[1,2]. Since indexing starts at 0, the expression refers to the item in the second row and third column of the array.

In NumPy, it is common to use notation, as in the preceding example, to index items in multidimensional arrays. This notation is possible because, in Python, arr[i, j, k, ...] can be used as shorthand for arr[(i, j, k, ...)]. Notice, however, that this behavior is defined in the ndarray class, and it is not possible, for example, to use multi-indices with pure Python lists of lists.

It is also valid to use negative indices to refer to positions from the end of the array. For instance, the last item of a row with index 2 can be referred to as a[2,-1].

As a general rule, a negative index -j refers to the item at position n-j, where n is the length of the axis along which the index is being used. This assumes that n-j is not negative.

Before moving on to the recipes in this section, let's define a function to help visualize arrays using the following code:

```
def dumpArray(arr):
    print('\n'.join([
        ' '.join(['{:02d}'.format(elem) for elem in row])
        for row in arr
    ]))
```

To make the examples concrete, we will use an array x, defined as follows:

```
m = n = 10
x = np.array([np.arange(n) + m*i for i in range(n)])
```

Using the code dumpArray(x), we can display the contents of the array x, as follows:

```
00 01 02 03 04 05 06 07 08 09
10 11 12 13 14 15 16 17 18 19
20 21 22 23 24 25 26 27 28 29
...
90 91 92 93 94 95 96 97 98 99
```

Notice that the array x is defined in such a way that its elements reflect their position in the array.

How to do it...

Let's learn how to access and select sub arrays. We will also get into the details of indexing.

Accessing sub arrays using slices

A *slice* is a reference to a subset of the elements in an array. Slices allow access to subranges of equally spaced items across each axis of the array. For example, the following code accesses a subset of elements on contiguous rows and columns of the array x:

```
y = x[2:4, 3:6]
dumpArray(y)
```

This will produce the following output:

```
23 24 25
33 34 35
```

Notice that, as is the case with ranges, the slice does not include its rightmost index. Thus, for instance, the slice 3:6 refers items at positions 3, 4, and 5.

NumPy follows the standard interface for Python slices, where i:j:k denotes a slice from i to j with step k. See the following link for the full documentation:
https://docs.python.org/3/library/stdtypes.html#sequence-types-list-tuple-range.

A slice can optionally include a step size, as exemplified by the following code:

```
y = x[1:5:2, 2:10:3]
dumpArray(y)
```

This code produces the output, shown as follows:

```
12 15 18
32 35 38
```

Using a negative increment, we can revert the order of the elements, as shown in the following example:

```
y = x[4:1:-1, 2]
```

This code will produce the following one-dimensional array:

```
array([42, 32, 22])
```

It is also valid to assign an array to a slice. For example, let's look at the following code, which demonstrates how to set a region of an array to zero:

```
xx = x.copy()
xx[2:-2, 2:-2] = np.zeros((6, 6), dtype=np.int64)
dumpArray(xx)
```

In this code, we first make a copy of the array x into xx using the `copy()` method. Then, we assign an array of zeros to the slice xx[2:-2, 2:-2], which has the effect setting a central rectangle of the array xx to zero, as follows:

```
00 01 02 03 04 05 06 07 08 09
10 11 12 13 14 15 16 17 18 19
20 21 00 00 00 00 00 00 28 29
30 31 00 00 00 00 00 00 38 39
40 41 00 00 00 00 00 00 48 49
50 51 00 00 00 00 00 00 58 59
60 61 00 00 00 00 00 00 68 69
70 71 00 00 00 00 00 00 78 79
80 81 82 83 84 85 86 87 88 89
90 91 92 93 94 95 96 97 98 99
```

The array being assigned to a slice must exactly match the dimensions of the slice.

In a slice `i:j:k`, it is possible to omit one or more of the indices `i`, `j`, and `k`. The slice is then interpreted as follows:

- If `i` is omitted, the start of the slice is assumed to be the beginning of the range
- If `j` is omitted, the end of the slice is assumed to be the end of the range
- If `k` is omitted, the increment in the slice is assumed to be one

It is actually possible to omit all indices in a slice expression, in which case the slice represents all items along the corresponding axis of the array. For example, the following code illustrates a typical idiom for selecting a row or column of an array:

```
y = x[:, 5]
y
```

This produces the following array:

```
array([ 5, 15, 25, 35, 45, 55, 65, 75, 85, 95])
```

 Notice that, apparently, NumPy returned a row vector, even though we requested a column from the original array. In fact, NumPy does not have a notion of row or column vectors. The output of this example is a vector with shape `(10,)`, that is, a tuple with the single element `10`. A one-dimensional NumPy array can be interpreted as either a row vector or a column vector, depending on the context. If we need to convert `y` to a two-dimensional array with shape `(1,10)`, we can use the code `y.shape=(1,10)`.

Finally, it is possible to refer to the whole array x using the `x[:]` slice expression. A word of caution is necessary, however. In pure Python, if `lst` is a list, the `lst[:]` expression causes the whole list to be copied. The same is not true with NumPy arrays. The following code illustrates the point:

```
xx = x.copy()
y = xx[:]
y[1,1] = 0
dumpArray(xx)
```

This will produce the following output:

```
00 01 02 03 04 05 06 07 08 09
10 00 12 13 14 15 16 17 18 19
20 21 22 23 24 25 26 27 28 29
...
```

Notice that entry (1,1) of the array xx was modified. This happens because the y=xx[:] statement creates a reference to the array xx, and no copying is effected. Thus, changes to the array y will actually change the array xx. NumPy adopts the policy of avoiding copying arrays unless a copy is absolutely necessary, or the user explicitly requests a copy. This is a wise decision, since arrays can be quite large, in which case copying becomes an expensive operation.

Having said that, let's compare the previous example with the following code:

```
xx = x.copy()
y[:] = xx
y[1,1] = 0
dumpArray(xx)
```

Running this code, the reader can verify that the array xx is *not* modified. This means that the y[:] = xx statement does indeed copy the array xx into y.

Whenever an assignment to a slice is made, NumPy assumes that a copy is requested.

Selecting subarrays using an index list

A slice allows for the selection of items from an array from equally spaced rows or columns. If a more general subarray is required, we can use lists as indices. This is illustrated with the following code:

```
y = x[[2, 3, 1], [0, 3, 5]]
```

This will assign array([20, 33, 15]) to the array y. The code behaves as if the lists [2, 3, 1] and [0,3,5] are zipped into the sequence of indices (2,0), (3,3), and (1,5). Also notice that the resulting array y is one-dimensional.

Combined with slices, list indexes provides a powerful way to extract items from an array. The following example demonstrates how to select a subset of the columns of an array:

```
y = x[:, [5, 2, -1]]
dumpArray(y)
```

This code has the effect of selecting columns with indices 5 and 2, as well as the last column from the array x, and produces the following output:

```
05 02 09
15 12 19
25 22 29
. . .
```

Notice that when using a list array this way, we can reorder the rows/columns of an array.

As a final example, let's look at how to extract an arbitrary (rectangular) subarray. This can be done with the following code:

```
y = x[[2, 0, 3],:][:,[-2, 1]]
dumpArray(y)
```

This produces the following output:

```
28 21
08 01
38 31
```

The preceding code selects the subarray in two steps:

- First, the [[2, 0, 3],:] index expression selects rows 2, 0, and 3 from the array x.
- Then, [:,[-2, 1]] selects columns -2 and 1 from the array. Notice that -2 refers to the second column from the end of the array.

Indexing with Boolean arrays

A Boolean array can be used to select a subset of the items in an array according to a given condition, as illustrated in the following code:

```
condition = x > 87
y = x[condition]
```

In this code, we first create the array condition, which is a Boolean array that holds a True value only at the positions that are larger than 87 in the array x. Next, we use the y=x[condition] statement to fetch the elements of x for which the condition is True. The overall effect is to generate an array that contains only the elements in x that are larger than 87, as follows:

```
array([88, 89, 90, 91, 92, 93, 94, 95, 96, 97, 98, 99])
```

Notice that, when using a Boolean array in this way, the output is always a one-dimensional.

Another common use of a Boolean array index is to select the rows or columns of an array according to a condition. The following example shows you how to select all rows from a two-dimensional array of integers that have an even sum:

```
x = np.array([[-1, 2, 4], [2, -3, 1], [2, 4,  0],
              [-5, 4, 3], [3,  1, 1], [2, 0, -4]])
row_sums = x.sum(axis=1)
y = x[row_sums % 2 == 0, :]
```

This is how this code works:

- `row_sums = x.sum(axis=1)` computes the row sums of the array x, storing the results in the `row_sums` array
- The `row_sums % 2 == 0` expression computes a one-dimensional Boolean array with the value `True` at the indices where `row_sums` has an even entry, and `False` otherwise
- Then, `x[row_sums % 2 == 0, :]` selects the rows of x for which the row sums are even.

After executing the preceding code, the following will be the contents of array y:

```
array([[ 2, -3,  1],
       [ 2,  4,  0],
       [-5,  4,  3],
       [ 2,  0, -4]])
```

Operations on arrays

NumPy defines a rich set of operations and functions for the `ndarray` type. NumPy defines the notion of a *universal function*, abbreviated as `ufunc`, which is a function object that can be applied to arbitrary arrays. Universal functions are objects of `ufunc` type, and NumPy provides a vast collection of built-in `ufunc` functions, covering all computations needed in scientific and data applications.

A `ufunc` is specialized towards the element by element application of a function. That is, if x is an array object, and f is a `ufunc`, the f(x) expression will apply the function f to every element of array x, and return a new object with the resulting values.

 A `ufunc` follows a strict functional protocol; applying the f function to x will never change the elements of the x arrays themselves, but return a new array with the values of f applied to each element of x. User-defined `ufunc` functions should also follow this protocol.

`ufunc` functions have many built-in characteristics, the two most important being:

- Broadcasting, which takes place when a function is applied to arrays that have different shapes. The most common scenario in which this happens is when an array has a single element along one of one axis. Broadcasting will then repeat that same element in an attempt to match the dimensions of other arrays appearing in the function call.
- Casting, which handles situations in which a function is being applied to arrays of different types.

The rules for broadcasting and casting can be complicated, but most of the time they behave as expected.

 `ufunc` functions are *vectorized*, that is, they operate on whole array dimensions, instead of on each element separately. In the background, the `ufunc` function calls C code, so that a loop along an array dimension is much faster than the corresponding Python code. It is recommended that computations in NumPy are done using `ufunc` functions whenever possible.

How to do it...

As we now know the gist of the operations that can be performed, we can take a look at how they are performed on the arrays.

Computing a function for all elements of an array

All scalar functions defined in NumPy follow the `ufunc` protocol, so that when applied to an array, they are applied to every array item. The following example shows you how to compute the sine function for all elements of a one-dimensional array:

```
x = np.pi * np.arange(0, 2, 0.25)
y = np.sin(x)
```

In this code, we first create an array x with equally spaced values. Then, with the `np.sin(x)` expression, we compute an array y containing the value of the sine function evaluated at each item of the array x.

It is important to notice that `ufunc` functions will operate on arrays with an arbitrary number of dimensions. For a unary `ufunc`, that is, a function that admits a single argument, the shape of the returned value is the same as the shape of the input.

Keep in mind that `ufunc` functions always operate on an element-by-element basis. For example, if A is a two-dimensional square array, the `np.exp(A)` expression computes the exponential of each element of A. This should not be confused with the exponential of a square matrix A, which is a different concept. Matrix operations are dealt with later in this book, in `Chapter 5`, *Matrices and Linear Algebra*.

Doing array operations

`ufunc` functions can also admit more than one argument. The most common example of this is array operations. For example, we can add two arrays with the following code:

```
x = np.array([1.2, -0.23, 3.4])
y = np.array([2.4, -1.7, -5.4])
z = x + y
```

This code uses the + operator to compute the component-wise sum, or vector sum, of the two one-dimensional arrays x and y, and stores the result in array z.

The reader may be somewhat surprised that we are discussing a binary operator here, since the topic of this section is `ufunc` functions, which are functions. Behind the scenes, NumPy calls `ufunc` to perform the binary operation. For example, the preceding computation can be done with the code z=np.add(x,y). NumPy uses the standard Python operator overloading the interface to map array operations into function calls.

The following example shows you how to add a value to all elements of an array:

```
x = np.array([1.2, -0.23, 3.4])
z = 1 + x
```

When interpreting this code, NumPy uses casting and broadcasting. Here are the details of what is going on in this example:

- NumPy realizes that, in the expression 1+x, 1 is an int and x is an ndarray of floats. Thus, the integer 1, viewed as a 1 x 1 array, is cast into an array of floats, and the whole expression is reinterpreted as np.array([1.0]) + x.
- In the np.array([1.0]) + x expression, the left operand has the shape (1,) and the right operand has shape (3,), so NumPy broadcasts the first operand, effectively reinterpreting the expression as np.array([1.0, 1.0, 1.0]) + x.
- At this point, the two operands have the same shape and the sum is computed by adding the arrays element by element.

Broadcasting is a powerful technique and, used correctly, provides code that is elegant and efficient. For example, let's use broadcasting to add a vector to each row of a two-dimensional array. This can be accomplished with the following code:

```
x = np.arange(12, dtype=np.float64).reshape(3, 4)
y = np.array([1, 2, 3, 4])
z = y + x
```

To understand what is happening here, let's analyze this code step by step.

The first line creates an array x with the numbers from 0 to 11 arranged in a 3 x 4 matrix:

```
array([[  0.,   1.,   2.,   3.],
       [  4.,   5.,   6.,   7.],
       [  8.,   9.,  10.,  11.]])
```

The next line defines y as the one-dimensional array array([1, 2, 3, 4]). Then, in the expression y+x, we are trying to add an array of shape (4,) with an array of shape (3,4). NumPy recognizes that the length of second dimension of x coincides with the length of y, so it broadcasts y into the array:

```
array([[1, 2, 3, 4],
       [1, 2, 3, 4],
       [1, 2, 3, 4]])
```

Finally, the broadcasted array is added to the array x, component by component. The overall effect is to add 1 to the first column of x, 2 to the second column, and so forth.

What if we want to add the values to the rows, instead of the columns, of the array? Notice that the following code will not work:

```
x = np.arange(12, dtype=np.float64).reshape(3, 4)
y = np.array([1, 2, 3])
z = y + x
```

This will raise an exception and NumPy will tell us that it cannot broadcast arrays together with shapes (3,) and (3,4). We can fix this with the following code:

```
x = np.arange(12, dtype=np.float64).reshape(3, 4)
y = np.array([[1], [2], [3]])
z = y + x
```

In this code, with the command y = np.array([[1], [2], [3]]), we make sure that y is an array with shape (3,1), so that y is broadcast to the following:

```
array([[1, 1, 1, 1],
       [2, 2, 2, 2],
       [3, 3, 3, 3]])
```

Adding this array to x effectively adds 1 to the first row of x, 2 to the second row, and 3 to the third row.

Computing matrix products

When used with NumPy arrays, the * operator represents element by element multiplication, so it cannot be used to compute the matrix product. The matrix product of two arrays is computed with the dot() method of the ndarray object, as demonstrated in the following code:

```
A = np.array([[2.3, 4.5, -3.1],
              [1.2, 3.2, 4.0]])
B = np.array([[ 2.1, -4.6, 0.5, 2.2],
              [-1.2, -3.0, 1.7, 3.2],
              [ 3.1, 2.6, 1.1, 2.3]])
C = A.dot(B)
```

In this code segment, we first define two arrays, A and B, with shapes, respectively, of (2,3) and (3,4). Notice that the number of columns of A is equal to the number of rows of B, so that the matrix product AB is well-defined. The matrix product is computed with the statement C = A.dot(B), which outputs the array as follows:

```
array([[-10.18, -32.14,   5.39,  12.33],
       [ 11.08,  -4.72,  10.44,  22.08]])
```

Notice that since A has 2 rows, and B has 4 columns, the product AB is an array of shape (2,4).

Instead of the using the dot() method as suggested previously, we could have used the module-level function np.dot(), writing instead C=np.dot(A,B). The effect is the same, and I tend to prefer using this method because it requires less typing.

The dot() method does the the right thing when computing matrix-vector and vector-matrix products. This is shown in the following examples, where we assume that the array A is the same as in the preceding example:

```
u = np.array([0.5, 2.3, -4.0])
v = np.array([3.2, -5.4])
w = A.dot(u)
z = v.dot(A)
```

We first define two one-dimensional arrays, u and v. Then, the w = A.dot(u) statement computes the matrix-vector product Au, and the z = v.dot(A) statement computes the vector-matrix product vA. Notice that, in these computations, u is interpreted as a column vector, and v is interpreted as a row vector. The dot() method in NumPy is smart enough to determine if a one-dimensional should be considered as a row or a column vector. This does not cause any issues, since if the dimensions of the arrays are not compatible, an exception will be raised.

NumPy defines a matrix class, for which the * operator computes matrix products. This class was included for pedagogical reasons only and its use is not recommended in production code. As a matter of fact, I recommend *never* using the matrix class.

Using masked arrays to represent invalid data

A common issue arising when working with data or doing computations is the presence of invalid values. Such values can arise either from data that is missing or as a result of operations that resulted in inconsistent data. We may also want to mask array elements that we know would raise errors in further computations.

Masked arrays in NumPy are supported by the `masked_array` class, which is defined in the `numpy.ma` module. To work with masked arrays, we need first to import this module, which can be done with the following code:

```
import numpy.ma as ma
```

How to do it...

In the following sections, we will get into the details of creating a masked array from the following:

- An explicit mask
- A condition

Creating a masked array from an explicit mask

This recipe applies when we know in advance the position of an array that should be masked. The following code illustrates this situation:

```
x = np.array([1.0, 3.2, -2.1, 4.3, 5.4])
xm = ma.masked_array(x, [0, 0, 1, 0, 0])
```

In the preceding code, we first define a one-dimensional array x with five items. Then, in the second line, we create the masked array xm, applying the mask [0, 0, 1, 0, 0] to the array x. This has the effect of masking the third entry of the array. By printing `repr(xm)`, we can glimpse how the array is represented, as shown in the following output:

```
masked_array(data = [1.0 3.2 -- 4.3 5.4],
             mask = [False False  True False False],
        fill_value = 1e+20)
```

There are these important points to notice in this output:

- The mask array is an array of Booleans. This is always the case; if an array of another type is given as the mask, it will be cast to a Boolean array.
- The mask is applied to the positions where the mask array is `True`.
- Masked values are replaced by a specific fill value, which is the floating point number `1e+20` in our example. If another fill value is desired, it can be passed in the `fill_value` argument of the constructor.

Creating a masked array from a condition

The `numpy.ma` module has a collection of functions geared towards common ways of constructing masked arrays. The following code illustrates how to create an array where all entries between 4 and 6 are masked:

```
x = np.array([
        [i+j for j in range(i, i+5)]
        for i in range(5)])
xm = ma.masked_inside(x, 4, 6)
```

In this code, we first generate the following array:

```
[[ 0  1  2  3  4]
 [ 2  3  4  5  6]
 [ 4  5  6  7  8]
 [ 6  7  8  9 10]
 [ 8  9 10 11 12]]
```

We then use the `ma.masked_inside(x, 4, 6)` function call to mask all elements of the array that are in the interval [4,6], resulting in the following masked array:

```
[[0 1 2 3 --]
 [2 3 -- -- --]
 [-- -- -- 7 8]
 [-- 7 8 9 10]
 [8 9 10 11 12]]
```

Notice that the endpoints are included in the interval specified by `masked_inside`.

The following table lists all functions available for the construction of specific masks and what they are used for:

Function	Operation
masked_equal(x, v)	Mask items in x equal to v
masked_greater(x, v)	Mask items in x greater than v
masked_greater_equal(x, v)	Mask items in x greater than or equal to v
masked_inside(x, v1, v2)	Mask items in x inside the interval [v1, v2]
masked_invalid(x)	Mask items in x that are NaN or infinite.
masked_less(x, v)	Mask items in x less than v
masked_less_equal(x, v)	Mask items in x less than or equal to v
masked_not_equal(x, v)	Mask items in x not equal to v
masked_outside(x, v1, v2)	Mask items in x outside interval [v1,v2]
masked_where(condition, v)	Mask items in x at positions corresponding to entries where the array condition is true

The masked_where function is the most flexible way to generate a masked array. For example, the following code shows how to mask an array of two-digit integers at the positions where the sum of the digits is strictly between 5 and 9:

```
def digit_sum(n):
    return n // 10 + n % 10
x = np.array(
        [[10*i+j for j in range(1,7)]
         for i in range(1,7)])
condition = (digit_sum(x) > 5) & (digit_sum(x) < 9)
xm = ma.masked_where(condition , x)
```

In this code block, we first define a digit_sum() function, which returns the sum of the digits of a two-digit integer. We then define the array x, which contains the following data:

```
[[11 12 13 14 15 16]
 [21 22 23 24 25 26]
 [31 32 33 34 35 36]
 [41 42 43 44 45 46]
 [51 52 53 54 55 56]
 [61 62 63 64 65 66]]
```

We then define the `condition` array, as follows:

```
condition = (digit_sum(x) > 5) & (digit_sum(x) < 9)
```

This array will have a `True` entry only at positions where the digit sum of elements of x are greater than 5 but less than 9. Then, the masked array is created with the following call:

```
xm = ma.masked_where(condition , x)
```

The resulting array xm will contain the data indicated, as follows:

```
[[11 12 13 14 -- --]
 [21 22 23 -- -- --]
 [31 32 -- -- -- 36]
 [41 -- -- -- 45 46]
 [-- -- -- 54 55 56]
 [-- -- 63 64 65 66]]
```

Using object arrays to store heterogeneous data

Up to this point, we only considered arrays that contained native elementary data types.

How to do it...

If we need an array containing heterogeneous data, we can create an array with arbitrary Python objects as elements, as shown in the following code:

```
x = np.array([2.5, 'a string', [2,4], {'a':0, 'b':1}])
```

This will result in an array with the `np.object` ;data type, as indicated in the output line as follows:

```
array([2.5, 'string', [2, 4], {'a': 0, 'b': 1}], dtype=object)
```

If the objects to be contained in the array are not known at construction time, we can create an empty array of objects with the following code:

```
x = np.empty((2,2), dtype=np.object)
```

The first argument, `(2,2)`, in the call to `empty()`, specifies the shape of the array, and `dtype=np.object` says that we want an array of objects. The resulting array is not really empty but has every entry set as equal to `None`. We can then assign arbitrary objects to the entries of `x`.

> In a NumPy array of objects, as in Python lists and tuples, the stored values are *references* to the objects, not copies of the objects themselves.

Defining, symbolically, a function operating on arrays

Anybody that has written numerical code will know that a common source of mistakes is the definition of functions that evaluate a complicated formula. One way around this problem is to use a package for symbolic computations, and we will take advantage of `sympy`, which is a compact Python symbolic package.

Getting ready

> If you are using Anaconda, `sympy` is already installed on your system. Otherwise, you will have to install it by using `pip3 install sympy`.

To see the full results of the following recipe, we assume that the reader is running Jupyter. Before getting started, run the following code in a Jupyter cell:

```
from sympy import *
init_printing(use_latex=True)
```

This will load all `sympy` functions and objects in the current namespace. This is usually not recommended, but in this example it is clearer to define the mathematical functions we will be using.

How to do it...

Let's now define the expression for the function we want to compute by running the following code in a Jupyter cell:

```
r, theta = symbols('r, theta')
x = r * cos(theta)
y = r * sin(theta)
x, y
```

In the first line in this code, we define the symbols r and theta. A symbol in sympy is used to represent the name of a mathematical variable.

Next, with the x = r * cos(theta) and y = r * sin(theta) assignments, we define two sympy expressions, x and y.

 Notice that the sin() and cos() functions that appear here are defined in sympy and are not the same as the np.sin() function and np.cos() from NumPy!

The last line in the cell, x, y, will print the x and y expressions in pretty format, according to the usual rules for typesetting mathematics.

The reader may have noticed that the preceding expressions compute the Cartesian coordinates for a point expressed in polar coordinates.

One weakness of symbolic code is that it tends to produce slow code, but sympy has a neat solution to this problem. We can create a NumPy ufunc from a set of sympy expressions, as follows:

```
polar_to_rectangular = lambdify((r, theta), (x, y), modules='numpy')
```

How it works...

The lambdify() sympy function returns a Python function that can be used to evaluate a set of expressions numerically. The arguments to lambdify() are described as follows:

- The first argument, (r, theta), is a tuple containing the symbols appearing in the expressions. These will map to the input arguments of the polar_to_rectangular() function.

- The second argument, `(x, y)`, specifies the expressions used to define the function. These must be `sympy` expressions that contain only the symbols given in the first argument, namely `r` and `theta` in this example.
- The `modules='numpy'` argument specifies that we want to use NumPy for numerical computations.

We can now use the `polar_to_rectangular()` function to convert arrays containing polar coordinates to rectangular coordinates, as follows:

```
r_values = np.array([2, 1.5, -3])
theta_values = np.array([np.pi/4, np.pi/2, -np.pi])
xvalues, yvalues = polar_to_rectangular(r_values, theta_values)
```

In this code, we first define two arrays, `r_values` and `theta_values`, containing, respectively, the values of `r` and `theta` for the points we want to convert. Then, we call the `polar_to_rectangular()` function to effect the conversion, storing the rectangular coordinates in the `xvalues` and `yvalues` arrays. Notice that, since `polar_to_rectangular()` is a `ufunc`, it can be applied to arrays with arbitrary shapes. In particular, it supports broadcasting, as illustrated in the following code:

```
r = 4
theta_values = np.linspace(0, 2*np.pi, 300)
xvalues, yvalues = polar_to_rectangular(r, theta_values)
```

In this code, the `polar_to_rectangular()` function is called with a scalar as the first argument, `r=4`, and an array as the second argument. The result is storing the rectangular coordinates of a circle of radius 4 in the `xvalues` and `yvalues` arrays.

3
Using Matplotlib to Create Graphs

This chapter introduces Matplotlib, a powerful graphical library for scientific and data plotting. We will cover the following recipes:

- Creating two-dimensional plots of functions and data
- Generating multiple plots in a single figure
- Setting line styles and markers
- Using different backends to display graphs
- Saving plots to disk
- Annotating graphs
- Generating histograms and box plots
- Creating three-dimensional plots
- Generating interactive displays in the Jupyter Notebook
- Object-oriented graph creation using `Artist` objects
- Creating a map with cartopy

Introduction

All work in scientific computing requires the display of graphical representations of data. Matplotlib is a full-fledged library for scientific and data plots.

Matplotlib can be used in two ways:

- Using functions in the `pyplot` module, which adopts a state-machine approach to plotting, where graphs are built step by step, with each command adding elements to the plot
- Using Matplotlib `Artist` objects to add elements to an `Axes` object directly

Most of this chapter concentrates on `pyplot`, which is the fastest way to build a graph from scratch.

Creating two-dimensional plots of functions and data

This recipe presents the basic kind of plot generated by Matplotlib: a two-dimensional display, with axes, where datasets and functional relationships are represented by lines. Besides the data being displayed, a good graph will contain a title (caption), axes labels, and, perhaps, a legend identifying each line in the plot.

Getting ready

Start Jupyter and run the following commands in an execution cell:

```
%matplotlib inline
import numpy as np
import matplotlib.pyplot as plt
```

How to do it...

Run the following code in a single Jupyter cell:

```
xvalues = np.linspace(-np.pi, np.pi)
yvalues1 = np.sin(xvalues)
yvalues2 = np.cos(xvalues)
plt.plot(xvalues, yvalues1, lw=2, color='red',
        label='sin(x)')
plt.plot(xvalues, yvalues2, lw=2, color='blue',
        label='cos(x)')
plt.title('Trigonometric Functions')
plt.xlabel('x')
```

```
plt.ylabel('sin(x), cos(x)')
plt.axhline(0, lw=0.5, color='black')
plt.axvline(0, lw=0.5, color='black')
plt.legend()
None
```

This code will insert the plot shown in the following screenshot into the Jupyter Notebook:

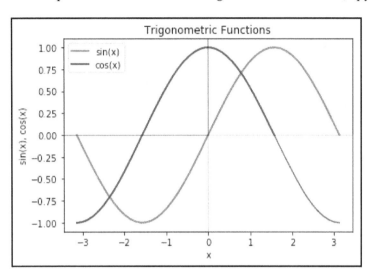

How it works...

We start by generating the data to be plotted, with the three following statements:

```
xvalues = np.linspace(-np.pi, np.pi, 300)
yvalues1 = np.sin(xvalues)
yvalues2 = np.cos(xvalues)
```

We first create an xvalues array, containing 300 equally spaced values between -π and π. We then compute the *sine* and *cosine* functions of the values in xvalues, storing the results in the yvalues1 and yvalues2 arrays. Next, we generate the first line plot with the following statement:

```
plt.plot(xvalues, yvalues1,
        lw=2, color='red',
        label='sin(x)')
```

The arguments to the `plot()` function are described as follows:

- `xvalues` and `yvalues1` are arrays containing, respectively, the *x* and *y* coordinates of the points to be plotted. These arrays must have the same length.
- The remaining arguments are formatting options. `lw` specifies the line width and `color` the line color. The `label` argument is used by the `legend()` function, discussed as follows.

The next line of code generates the second line plot and is similar to the one explained previously. After the line plots are defined, we set the title for the plot and the legends for the axes with the following commands:

```
plt.title('Trigonometric Functions')
plt.xlabel('x')
plt.ylabel('sin(x), cos(x)')
```

We now generate axis lines with the following statements:

```
plt.axhline(0, lw=0.5, color='black')
plt.axvline(0, lw=0.5, color='black')
```

The first arguments in `axhline()` and `axvline()` are the locations of the axis lines and the options specify the line width and color.

We then add a legend for the plot with the following statement:

```
plt.legend()
```

Matplotlib tries to place the legend intelligently, so that it does not interfere with the plot. In the legend, one item is being generated by each call to the `plot()` function and the text for each legend is specified in the `label` option of the `plot()` function.

Generating multiple plots in a single figure

Wouldn't it be interesting to know how to generate multiple plots in a single figure? Well, let's get started with that.

Getting ready

Start Jupyter and run the following three commands in an execution cell:

```
%matplotlib inline
import numpy as np
import matplotlib.pyplot as plt
```

How to do it...

Run the following commands in a Jupyter cell:

```
plt.figure(figsize=(6,6))
xvalues = np.linspace(-2, 2, 100)
plt.subplot(2, 2, 1)
yvalues = xvalues
plt.plot(xvalues, yvalues, color='blue')
plt.xlabel('$x$')
plt.ylabel('$x$')
plt.subplot(2, 2, 2)
yvalues = xvalues ** 2
plt.plot(xvalues, yvalues, color='green')
plt.xlabel('$x$')
plt.ylabel('$x^2$')
plt.subplot(2, 2, 3)
yvalues = xvalues ** 3
plt.plot(xvalues, yvalues, color='red')
plt.xlabel('$x$')
plt.ylabel('$x^3$')
plt.subplot(2, 2, 4)
yvalues = xvalues ** 4
plt.plot(xvalues, yvalues, color='black')
plt.xlabel('$x$')
plt.ylabel('$x^3$')
plt.suptitle('Polynomial Functions')
plt.tight_layout()
plt.subplots_adjust(top=0.90)
None
```

Running this code will produce results like those in the following screenshot:

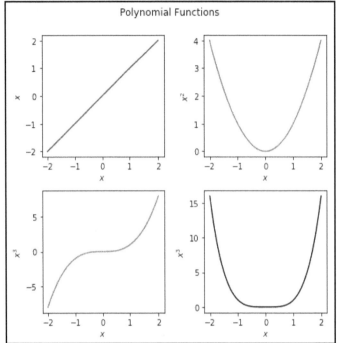

How it works...

To start the plotting constructions, we use the `figure()` function, as shown in the following line of code:

```
plt.figure(figsize=(6,6))
```

The main purpose of this call is to set the figure size, which needs adjustment, since we plan to make several plots in the same figure. After creating the figure, we add four plots with code, as demonstrated in the following segment:

```
plt.subplot(2, 2, 3)
yvalues = xvalues ** 3
plt.plot(xvalues, yvalues, color='red')
plt.xlabel('$x$')
plt.ylabel('$x^3$')
```

In the first line, the `plt.subplot(2, 2, 3)` call tells `pyplot` that we want to organize the plots in a two-by-two layout, that is, in two rows and two columns. The last argument specifies that all following plotting commands should apply to the third plot in the array. Individual plots are numbered, starting with the value `1` and counting across the rows and columns of the plot layout.

We then generate the line plot with the following statements:

```
yvalues = xvalues ** 3
plt.plot(xvalues, yvalues, color='red')
```

The first line of the preceding code computes the `yvalues` array, and the second draws the corresponding graph. Notice that we must set options such as line color individually for each subplot.

After the line is plotted, we use the `xlabel()` and `ylabel()` functions to create labels for the axes. Notice that these have to be set up for each individual subplot too.

After creating the subplots, we explain the subplots:

- `plt.suptitle('Polynomial Functions')` sets a common title for all subplots
- `plt.tight_layout()` adjusts the area taken by each subplot, so that axes' legends do not overlap
- `plt.subplots_adjust(top=0.90)` adjusts the overall area taken by the plots, so that the title displays correctly

Setting line styles and markers

Good displays of data should make it easy to identify what data is being represented by each graph component. There are several ways to do that, but using line colors, styles, and markers goes a long way to help readers identify what quantity is represented by each line.

Getting ready

Start Jupyter and run the following three commands in an execution cell:

```
%matplotlib inline
import numpy as np
import matplotlib.pyplot as plt
```

How to do it...

Run the following code in a Jupyter code cell:

```
plt.figure(figsize=(8,4))
plt.subplot(1, 2, 1)
xvalues = np.linspace(0, 1, 200)
plt.plot(xvalues, ls = '-', color='green')
plt.plot(xvalues + 1, ls = '--', color='green')
plt.plot(xvalues + 2, ls = ':', color='green')
plt.plot(xvalues + 3, ls = '-.', color='green')
plt.subplot(1, 2, 2)
xvalues = np.linspace(0, 1, 6)
plt.plot(xvalues, marker = 'o', ls='', color='blue')
plt.plot(xvalues + 1, marker = '^', ls='', color='blue')
plt.plot(xvalues + 2, marker = '*', ls='--', color='blue')
plt.plot(xvalues + 3, marker = '>', ls=':', color='blue',
         mec='black', mfc='red', ms=15)
None
```

This will produce the following output:

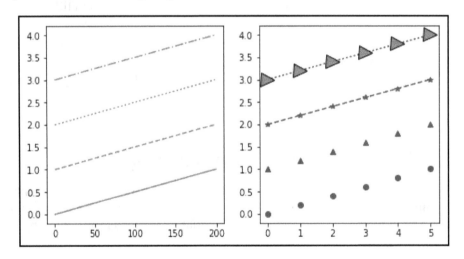

How it works...

The line plots with different line styles are produced with the following code:

```
plt.plot(xvalues, ls = '-', color='green')
plt.plot(xvalues + 1, ls = '--', color='green')
plt.plot(xvalues + 2, ls = ':', color='green')
plt.plot(xvalues + 3, ls = '-.', color='green')
```

The line style is set with the `ls` option, which accepts the values shown in the following table:

Option	Line style
`'-'` or `'solid'`	Solid line
`'--'` or `'dashed'`	Dashed line
`'-.'` or `'dashdot'`	Dashed-dotted line
`':'` or `'dotted'`	Dotted line
`None,` `' '` or `''`	Blank line

In the second subplot, we demonstrate the use of markers with the following code:

```
plt.plot(xvalues, marker='o', ls='', color='blue')
plt.plot(xvalues + 1, marker='^', ls='', color='blue')
plt.plot(xvalues + 2, marker='*', ls='--', color='blue')
plt.plot(xvalues + 3, marker='>', ls=':', color='blue',
         mec='black', mfc='red', ms=15)
```

Markers are specified with the `marker` option and a wide variety of marker shapes are supported. The possible values for the marker option are listed at `https://matplotlib.org/api/markers_api.html#module-matplotlib.markers`.

Markers and line styles are independent and must be set with different options. In the first two preceding examples, we use `ls=''` to specify a *blank* line, that is, no connecting lines are drawn between the points. In the next two lines of code, we use, respectively, a dashed line and a dotted line.

In the last example, we used the `mec='black'`, `mfc='red'`, and `ms=15` options to, respectively, set the marker edge color to black, the marker face color to red, and the marker size to 15.

Using different backends to display graphs

Matplotlib can produce high-quality graphs targeting many different platforms and display technologies. To achieve this in an efficient and transparent way, Matplotlib developers introduced the notion of a *backend*. A backend is code that is designed to render Matplotlib graphs in a particular graphing environment. Users of Matplotlib are mostly insulated from the details of the actual rendering, but some devices and operating systems might require the selection of the appropriate backend.

Getting ready

In this example, we will write our code in a script, so you will need to open your text editor.

How to do it...

Enter the following code in your text editor:

```python
import numpy as np
import matplotlib
matplotlib.use('Qt5Agg')
import matplotlib.pyplot as plt

f = lambda x: 1/((x-3)**2+.01) + 1/((x-9)**2+.04) - 6
xvalues = np.linspace(2, 10, 300)
yvalues = f(xvalues)

plt.plot(xvalues, yvalues)
plt.xlabel('$x$')
plt.ylabel('$f(x)$')
plt.title('The "humps" function')

plt.show()

print('Done plotting.')
```

Save the script to disk with the name `backend_test.py` and run it with the following command line:

```
python3 backend_test.py
```

Running the script will open a window displaying the plot shown in the following screenshot:

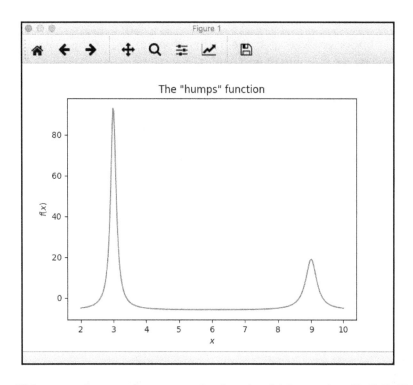

This example uses the Qt5Agg backend, which requires PyQt5. If you are using Anaconda or followed the installation instructions in Chapter 1, *Getting to Know the Tools*, you already have PyQt5. Otherwise, to make this example work, you will need to install it by running pip install pyqt5.

How it works...

The script starts by importing the required libraries with the following code:

```
import numpy as np
import matplotlib
matplotlib.use('Qt5Agg')
import matplotlib.pyplot as plt
```

The order or the commands is important here. We first import Matplotlib and then issue the `matplotlib.use(Qt5Agg)` call to select the `Qt5Agg` backend. After selecting the backend, we import `pyplot`.

 It is important to choose the backend before importing `pyplot`. Otherwise, the default backend, `Agg`, will be selected. The `Agg` backend is non-interactive and is used for graphs to be stored on disk only.

The next lines of code generate the data and create the plot. As the plot is defined, no output is actually generated. To display the plot, we need to call the `show()` function, as shown in the next line of code, reproduced as follows:

```
plt.show()
```

This will direct the backend to render the plot, which is displayed in a new window.

Finally, we tell Python to print a message to standard output with the following statement:

```
print('Done plotting.')
```

Notice that, if you are running the code as directed in the example, nothing is printed at this point. The `Qt5Agg` backend is *blocking*, which means that it blocks execution to the script. Closing the plot window will release the block, and the script will continue, print the `Done plotting.` message, and exit.

The following is a partial list of the available backends, including only backends compatible with Python 3. Notice that each backend has its own requirements, that must be installed using `pip`.

Backend	Description	Interactive	File type
GTK3Agg	Rendering in a GTK 3.x canvas. Requires `PyGObject` and `pycairo`.	Yes	None
GTK3Cairo	Rendering in a GTK 3.x canvas. Requires `PyGObject` and `pycairo`.	Yes	None
WXAgg	Rendering to a wxWidgets canvas. Requires `wxPython`.	Yes	None
TkAgg	`Agg` rendering to a Tk canvas. Requires `TkInter`.	Yes	None
Qt4Agg	Rendering to a Qt4 canvas. Requires PyQt4 or PySide.	Yes	None
Qt5Agg	Rendering to a Qt5 canvas. Requires PyQt5.	Yes	None
macosx	Rendering to a Cocoa window in macOS. Graph window is non-blocking.	Yes	None
Agg	Raster graphics using `Agg` (anti-grain geometry).	No	png
PS	Postscript format.	No	ps or eps
PDF	Portable document format.	No	pdf

| SVG | Scaled vector graphics. | No | svg |
| cairo | Vector graphics with Cairo graphics. | No | png, ps, pdf, svg; |

In the preceding table, interactive backends open a window in the user's desktop, which allows the graph to be modified and blocks the execution of the script that generated the graph (with the exception of the `macosx` backend). Non-interactive backends are designed for saving graphs to disk, which is discussed in the next recipe.

Saving plots to disk

It is often necessary to save the results of plots in permanent storage. The reasons for that include exporting plots to other software, sharing the graphs, and creating posters or presentations. Another reason to save graphs is to preserve different versions of the same plot.

Getting ready

Start Jupyter and run the following three commands in an execution cell:

```
%matplotlib inline
import numpy as np
import matplotlib.pyplot as plt
```

How to do it...

To save a plot to disk, we use the `savefig()` function, as illustrated in the following code segment:

```
plt.figure(figsize=(4,4))
tvalues = np.linspace(-np.pi, np.pi, 500)
xvalues = 2*np.cos(tvalues)-np.cos(2*tvalues)
yvalues = 2*np.sin(tvalues)-np.sin(2*tvalues)
plt.plot(xvalues, yvalues, color='brown')
plt.axhline(0, color='black', lw=1)
plt.axvline(0, color='black', lw=1)
plt.title('Cardioid')
plt.savefig('cardioid.png')
None
```

How it works...

In this code, we first create a new figure and set its size. Then, the `tvalues`, `xvalues`, and `yvalues` arrays are filled with the values to be displayed, which, in this case, represent a parametric plot of a cardioid. Finally, after the line plots and other plot elements are added, we issue the following function call:

```
plt.savefig('cardioid.png')
```

This will output the graph to the `cardioid.png` file, which is created if it does not exist.

> Be aware that, if the file already exists, it will be overwritten! In general, this is not a problem, since we can always rebuild the graph in Matplotlib. However, to compare different versions of the plot, remember to choose a different name each time the plot is saved.

The format of the output file is determined from the extension in the given filename. In our example, we use the `.png` extension, which generates a file in the Portable Network Graphics format, which is a common format for images in web pages. If the `cardioid.png` file is opened in a previewer, the following figure will be displayed:

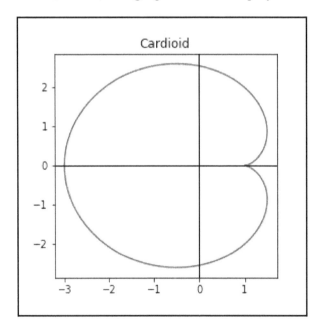

The `savefig()` function has several useful options. Some of the most useful are listed in the following table:

Option	Description
dpi	Image resolution in dots per inch. Set dpi='figure' to use the same resolution as the figure object.
facecolor, edgecolor	The colors of the bounding rectangle where the figure is embedded.
format	The graphic output format, if not specified in the file extension.
transparent	True or False. If True, the bounding rectangle is transparent.
box_inches	The size of the bounding rectangle in inches. Set bbox_inches='tight' to use the smallest rectangle that fits the figure.
pad_inches	The amount of padding when using a tight bounding rectangle.

When saving a figure to disk, Matplotlib adds a *bounding rectangle* behind the graph. By default, this rectangle is white. The color of the background rectangle can be controlled with the `facecolor` and `edgecolor` options.

In some situations, the size automatically chosen for the bounding rectangle is excessive. We can then use the `bbox_inches='tight'` option, shown as follows, to improve the appearance of the graph:

```
plt.savefig('cardioid_tight.png', bbox_inches='tight')
```

Another useful set of options is shown as follows, which can be used to change the background color of a figure:

```
plt.savefig('cardioid_bgyellow.png',
            transparent=True, facecolor='lightyellow')
```

Notice that it is necessary to set `transparent=True`, otherwise only the bounding rectangle will be colored, resulting in a yellow margin, instead of the desired effect of a yellow background.

By default, the image is produced with a resolution appropriate for a screen display. Images that are intended to be printed, especially in large media such as a poster, should use the `dpi` option, as illustrated in the following command:

```
plt.savefig('cardioid_highres.png', dpi=300)
```

In this code, we set `dpi=300`, which is the recommended resolution for printed images.

The `format` option can be used to set the image format of the output, if it was not specified in the file extension. The formats supported depend on the active backend, but, in general, `.png`, `.pdf`, `.jpg`, and `.svg` are supported.

Annotating graphs

It is often convenient to annotate plots to indicate relevant features, or to add text with commentary to the graph. Matplotlib provides two main functions to add these features to a plot:

- `annotate()` creates an annotation associated to a specific point in a graph, with an optional arrow pointing to the relevant value
- `text()` adds generic text to a graph

The next recipe shows how to add the two kinds of annotation.

Getting ready

Start Jupyter and run the following three commands in an execution cell:

```
%matplotlib inline
import numpy as np
import matplotlib.pyplot as plt
```

How to do it...

Run the following code in an execution cell:

```
from scipy.stats import gamma
distr = gamma(2.0)
xvalues = np.linspace(0, 10, 100)
yvalues = distr.pdf(xvalues)
plt.plot(xvalues, yvalues, color='orange')
plt.title('Mean and median')
plt.xlabel('$x$')
plt.ylabel('$f(x)$')
xmean = distr.mean()
ymean = distr.pdf(xmean)
```

```
xmed = distr.median()
ymed = distr.pdf(xmed)
aprops = dict(arrowstyle = '->')
plt.annotate('Median={:3.2f}'.format(xmed),
             (xmed, ymed),
             (xmed+1, ymed+0.03),
             arrowprops=aprops)
plt.annotate('Mean={:3.2f}'.format(xmean),
             (xmean, ymean),
             (xmean+1, ymean+0.03),
             arrowprops=aprops)
comment = '''In a skewed distribution,
mean and median usually
have different values.'''
plt.text(4.5, 0.15, comment, fontsize=12, color='blue')
None
```

Running this code creates the graph of a skewed probability distribution, with annotations showing where the **Mean** and **Median** are located, as shown in the following screenshot:

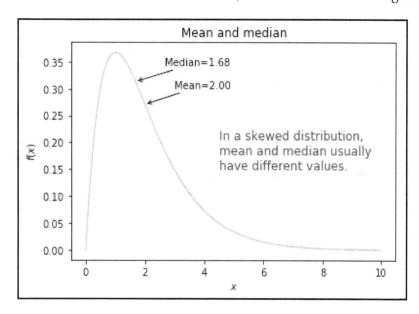

How it works...

The code starts by generating a plot of the probability mass function of a Gamma distribution, which is a well-known example of skewed distribution. After importing the `gamma` object from `scipy.stats`, the distribution is created with `distr = gamma(2.0)`. The value of `2.0` for the parameter is chosen to generate a graph with the desired shape. The points defining the plot are then computed in the `xvalues` and `yvalues` arrays and the `plot()` function is called to create the line plot. After that, we add a title and labels for the axes with the `title()`, `xlabel()`, and `ylabel()` functions, respectively.

After the line plot is set up, we start working on the annotations. The first step is to compute the mean and median of the distribution with the following code:

```
xmean = distr.mean()
ymean = distr.pdf(xmean)
xmed = distr.median()
ymed = distr.pdf(xmed)
```

To compute the mean and median we use, respectively, the `mean()` and `median()` methods of the `distr` object. We also compute the *y* coordinates corresponding to the mean and median and store their values in the `ymean` and `ymed` variables.

The next code line creates a dictionary named `aprops`:

```
aprops = dict(arrowstyle = '->')
```

This dictionary will be used to set the arrow properties in the annotation. We are using a simple arrow style with the `arrowstyle = '->'` option. Matplotlib has an extensive set of options for formatting fancy arrows, represented by objects of `FancyArrowPatch` type, documented at `https://matplotlib.org/2.0.2/api/patches_api.html#matplotlib.patches.FancyArrowPatch`.

We are now ready to add the annotations. To annotate the point in the graph corresponding to the median, we use the following code:

```
plt.annotate('Median={:3.2f}'.format(xmed),
             (xmed, ymed),
             (xmed+1, ymed+0.03),
             arrowprops=aprops)
```

The call to `annotate()` in the previous code uses the following arguments:

- The first argument is a string, containing the text to be displayed. In our case, the string displays information about the median of the distribution.
- The second argument, `(xmed, ymed)`, specifies the point being annotated. This corresponds to the tip of the arrow.
- The third argument, `(xmed+1, ymed+0.03)`, indicates the position of the text. We position the text a little bit to the right of and above the point being annotated.
- The last argument, `arrowprops=aprops`, specifies the arrow properties, using the `aprops` dictionary discussed previously.

Next, we make another call to `annotate()`, with analogous arguments, to add an annotation to the point corresponding to the mean of the distribution.

The final step in the graph construction is to add the commentary text with the following code:

```
comment = '''In a skewed distribution,
mean and median usually
have different values.'''
plt.text(4.5, 0.15, comment, fontsize=12, color='blue')
```

We first store the multiline comment string in the comment variable. Then, the text is added to the figure with a call to the `text()` function, using the following arguments:

- The first two arguments, `4.5` and `0.15`, specify the *x* and *y* coordinates of the lower-left point of the text.
- The second argument, `comment`, is the text to be added.
- All other arguments are interpreted as formatting options for the text. We use `fontsize=12` to select a larger font, and `color='blue'` to display the comment in blue.

Generating histograms and box plots

Matplotlib supports the creation of a variety of displays of data. In this recipe, we will demonstrate how to use two popular graphs representing data variability: histograms and box plots (also known as box-and-whisker plots). We will present a comparison between the distribution of heights in the male and female populations. To make the example self-contained, instead of using real data, we will simulate a population with the known distribution of heights for males and females.

Getting ready

Start Jupyter and run the following three commands in an execution cell:

```
%matplotlib inline
import numpy as np
import matplotlib.pyplot as plt
```

How to do it...

Run the following code in a Jupyter code cell:

```
from scipy.stats import norm
mmean, msdev = 70, 4.0
fmean, fsdev = 65, 3.5
mdist = norm(mmean, scale=msdev)
fdist = norm(fmean, scale=fsdev)
nm, nf = 2000, 1500
mdata = mdist.rvs(size=nm)
fdata = fdist.rvs(size=nf)
plt.figure(figsize=(12, 4))
plt.subplot(1,2,1)
plt.hist([mdata, fdata], bins=20,
         label=['Males', 'Females'])
plt.legend()
plt.xlabel('Height (inches)')
plt.ylabel('Frequency')
plt.subplot(1,2,2)
plt.boxplot([mdata, fdata], patch_artist=True,
            labels=['Males', 'Females'])
None
```

Running this code will generate the following graph:

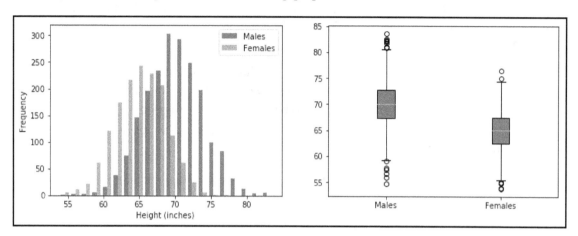

The two graphs clearly display the characteristics of a normal distribution, with heights for the male population having a larger mean and standard deviation. The box plots also display the outliers in the distribution, according to the 1.5 IQR rule.

How it works...

We start this example by simulating populations of males and females, according to the known parameters for the distribution of heights in each subpopulation. Since heights are normally distributed, we use the `norm` object from the `scipy.stats` module to generate the data. For example, the male population is generated with the following statement:

```
mdata = mdist.rvs(size=nm)
```

After the data is generated, we issue the `plt.figure(figsize=(12, 4))` function call, which sets the figure dimensions. The histogram is then generated by executing the following code:

```
plt.subplot(1,2,1)
plt.hist([mdata, fdata], bins=20,
         label=['Males', 'Females'])
plt.legend()
plt.xlabel('Height (inches)')
plt.ylabel('Frequency')
```

Since we want the plots to appear side by side, we use the `plt.subplot(1,2,1)` command, which specifies a layout with one row and two columns, and sets the current target graph to be in the first one. The `hist()` function is then called with the following arguments:

- `[mdata, fdata]` is a list containing the datasets to be displayed in the histogram. By default, multiple datasets are displayed side by side.
- `bins=20` sets the number of bins to be used in the histogram.
- `label=['Males', 'Females']` specifies the labels to be used for each dataset in the legend.

Next, calling `plt.legend()` causes the legend to be added to the graph and we also set the labels in the two axes with calls to `xlabel()` and `ylabel()`.

The next statements plot the box plot with the following lines of code:

```
plt.subplot(1,2,2)
plt.boxplot([mdata, fdata], patch_artist=True,
            labels=['Males', 'Females'])
```

After selecting the subplot, we call the `boxplot()` function to generate the plot, with the two datasets given in the first argument. The `patch_artist=True` option directs Matplotlib to use the `Artist` interface, which produces a nicer plot. Finally, `labels=['Males', 'Females']` sets the labels along the *x* axis for each of the box plots.

Creating three-dimensional plots

Matplotlib offers several different ways to visualize three-dimensional data. In this recipe, we will demonstrate the following methods:

- Drawing surfaces plots
- Drawing two-dimensional contour plots
- Using color maps and color bars

Getting ready

Start Jupyter and run the following three commands in an execution cell:

```
%matplotlib inline
import numpy as np
import matplotlib.pyplot as plt
```

How to do it...

Run the following code in a Jupyter code cell:

```
from mpl_toolkits.mplot3d import Axes3D
from matplotlib import cm
f = lambda x,y: x**3 - 3*x*y**2
fig = plt.figure(figsize=(12,6))
ax = fig.add_subplot(1,2,1,projection='3d')
xvalues = np.linspace(-2,2,100)
yvalues = np.linspace(-2,2,100)
xgrid, ygrid = np.meshgrid(xvalues, yvalues)
zvalues = f(xgrid, ygrid)
surf = ax.plot_surface(xgrid, ygrid, zvalues,
                       rstride=5, cstride=5,
                       linewidth=0, cmap=cm.plasma)
ax = fig.add_subplot(1,2,2)
plt.contourf(xgrid, ygrid, zvalues, 30,
             cmap=cm.plasma)
fig.colorbar(surf, aspect=18)
plt.tight_layout()
None
```

Running this code will produce a plot of the *monkey saddle* surface, which is a famous example of a surface with a non-standard critical point. The displayed graph is shown in the following screenshot:

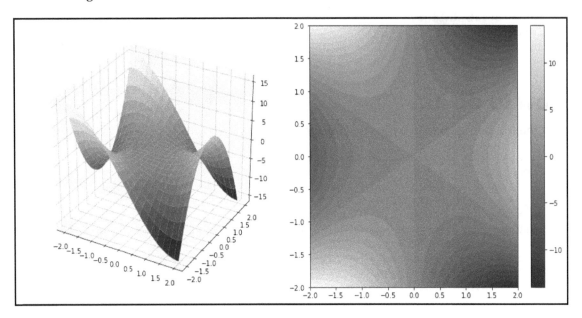

How it works...

We start by importing the `Axes3D` class from the `mpl_toolkits.mplot3d` library, which is the Matplotlib object used for creating three-dimensional plots. We also import the `cm` class, which represents a color map. We then define a function to be plotted, with the following line of code:

```
f = lambda x,y: x**3 - 3*x*y**2
```

The next step is to define the `Figure` object and an `Axes` object with a 3D projection, as done in the following lines of code:

```
fig = plt.figure(figsize=(12,6))
ax = fig.add_subplot(1,2,1,projection='3d')
```

Notice that the approach used here is somewhat different than the other recipes in this chapter. We are assigning the output of the `figure()` function call to the `fig` variable and then adding the subplot by calling the `add_subplot()` method from the `fig` object.

This is the recommended method of creating a three-dimensional plot in the most recent version of Matplotlib. Even in the case of a single plot, the `add_subplot()` method should be used, in which case the command would be `ax = fig.add_subplot(1,1,1,projection='3d')`.

The next few lines of code, shown as follows, compute the data for the plot:

```
xvalues = np.linspace(-2,2,100)
yvalues = np.linspace(-2,2,100)
xgrid, ygrid = np.meshgrid(xvalues, yvalues)
zvalues = f(xgrid, ygrid)
```

The most important feature of this code is the call to `meshgrid()`. This is a NumPy convenience function that constructs grids suitable for three-dimensional surface plots. To understand how this function works, run the following code:

```
xvec = np.arange(0, 4)
yvec = np.arange(0, 3)
xgrid, ygrid = np.meshgrid(xvec, yvec)
```

After running this code, the `xgrid` array will contain the following values:

```
array([[0, 1, 2, 3],
       [0, 1, 2, 3],
       [0, 1, 2, 3]])
```

The `ygrid` array will contain the following values:

```
array([[0, 0, 0, 0],
       [1, 1, 1, 1],
       [2, 2, 2, 2]])
```

Notice that the two arrays have the same dimensions. Each grid point is represented by a pair of the `(xgrid[i,j],ygrid[i,j])` type. This convention makes the computation of a vectorized function on a grid easy and efficient, with the `f(xgrid, ygrid)` expression.

The next step is to generate the surface plot, which is done with the following function call:

```
surf = ax.plot_surface(xgrid, ygrid, zvalues,
                       rstride=5, cstride=5,
                       linewidth=0, cmap=cm.plasma)
```

The first three arguments, xgrid, ygrid, and zvalues, specify the data to be plotted. We then use the rstride and cstride options to select a subset of the grid points. Notice that the xvalues and yvalues arrays both have length 100, so that xgrid and ygrid will have 10,000 entries each. Using all grid points would be inefficient and produce a poor plot from the visualization point of view. Thus, we set rstride=5 and cstride=5, which results in a plot containing every fifth point across each row and column of the grid.

The next option, linewidth=0, sets the line width of the plot to zero, preventing the display of a wireframe. The final argument, cmap=cm.plasma, specifies the color map for the plot. We use the cm.plasma color map, which has the effect of plotting higher functional values with a *hotter* color. Matplotlib offer as large number of built-in color maps, listed at https://matplotlib.org/examples/color/colormaps_reference.html.

Next, we add the filled contour plot with the following code:

```
ax = fig.add_subplot(1,2,2)
ax.contourf(xgrid, ygrid, zvalues, 30,
            cmap=cm.plasma)
```

Notice that, when selecting the subplot, we do not specify the projection option, which is not necessary for two-dimensional plots. The contour plot is generated with the contourf() method. The first three arguments, xgrid, ygrid, zvalues, specify the data points, and the fourth argument, 30, sets the number of contours. Finally, we set the color map to be the same one used for the surface plot.

The final component of the plot is a color bar, which provides a visual representation of the value associated with each color in the plot, with the fig.colorbar(surf, aspect=18) method call.

Notice that we have to specify in the first argument which plot the color bar is associated to. The aspect=18 option is used to adjust the aspect ratio of the bar. Larger values will result in a narrower bar.

To finish the plot, we call the tight_layout() function. This adjusts the sizes of each plot, so that axis labels are displayed correctly.

Generating interactive displays in the Jupyter Notebook

In many situations, information changes dynamically, and it is desirable to present the changes graphically. In this recipe, we will discuss the creation of *interacts* in the Jupyter Notebook, which can be used to generate dynamic displays based on function evaluations. As an example, we will demonstrate how to create an interactive demonstration of the *beats* phenomenon from acoustics.

Getting ready

Start Jupyter and run the following three commands in an execution cell:

```
%matplotlib inline
import numpy as np
import matplotlib.pyplot as plt
```

How to do it...

We start by defining a function that plots a superposition of waves, with the following code:

```
xvalues = np.linspace(0, 40, 500)
def plot_function(ratio=0.1):
    plt.figure(figsize=(10,3))
    yvalues = np.cos(2*np.pi*xvalues) + \
            np.cos(2*np.pi*ratio*xvalues)
    plt.plot(xvalues, yvalues,
            color='green', lw=2)
    plt.axis([0, 40, -2.1, 2.1])
    plt.show()
```

Calling `plot_function()` by itself produces a static plot of a wave function. The `ratio` argument is used to specify the ratio between the frequencies of the superimposed waves.

To create an interactive display, simply execute the following in a Jupyter code cell:

```
from ipywidgets import interact
interact(plot_function, ratio=(0.1, 1.1, 0.01))
None
```

This will generate a display similar to the one shown in the following screenshot:

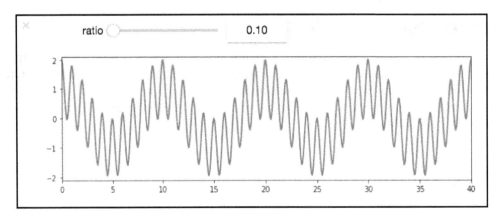

Notice the presence of a slider interface element (also known as a **widget**). By moving the slider, we can update the value of the `ratio` parameter passed to `plot_function()`. As we change the ratio towards values closer to **1**, we get oscillations that display *beats*, as shown in the following screenshot:

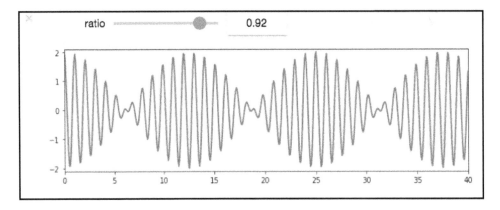

How it works...

We start by defining a function that creates a plot, naming it `plot_function()`. This function takes one argument, `ratio`, which is the variable that the user will control interactively. Notice that we specify a default value for the `ratio` argument, with the `ratio=0.1` option. This default value is used as the initial value of the interactive variable.

In the body of `plot_function()`, we construct a Matplotlib graph using the usual commands. Notice, however, that we add a call to `show()` at the end of the function. This is necessary to make the function work correctly with Jupyter's display system.

The interact is set up with the following call:

```
interact(plot_function, ratio=(0.1, 1.1, 0.01))
```

The first argument to `interact()` is the function that generates the plot, `plot_function`. The second argument, `ratio=(0.1, 1.1, 0.01)`, specifies the details of the slider control for the `ratio` variable. In this case, the values of `ratio` will be constrained between `0.1` and `1.1`, with a step of `0.01` between successive values.

A Matplotlib interact accepts a varied collection of widgets, depending on the type of variable being used for the interaction. The following table summarizes the kind of control associated with each type of keyword argument:

Keyword argument type	Example	Associated widget
`bool`	`bvar=True`	Checkbox
`int`	`ivar=(1, 10)`	Integer slider
`float`	`bvar=(0.0,2.0,0.1)`	Float slider
`str`	`svar='Hello, world!'`	Text
`list` of `str`	`lvar = ['a', 'b', 'c']`	Dropdown

The `ipywidgets` module supports the creation of sophisticated user interfaces, with several different kinds of widgets and layout options. The interested reader is encouraged to explore the documentation of the package, by visiting `http://ipywidgets.readthedocs.io/en/latest/user_guide.html`.

Object-oriented graph creation using Artist objects

Up to this point, we have used Matplotlib's high-level functions, such as `plot()`, to create graphs. In this section, we will show how to use objects of type `Artist` to draw arbitrary shapes in a plot.

In Matplotlib, every plot element is an object from the `Artist` class. There are two kinds of `Artist` objects:

- **Containers** are objects designed to contain other graphic elements. The most important types of container are `Figure` and `Axes` objects.
- **Primitives** represent elements that can be directly drawn in a container. These include objects of type `Line2D`, `Patch`, and `Text`, for example.

In the next recipe, we will see how to use Matplotlib's `Artist` interface to draw general shapes in a figure.

Getting ready

Start Jupyter and run the following three commands in an execution cell:

```
%matplotlib inline
import numpy as np
import matplotlib.pyplot as plt
```

How to do it...

Enter the following code in a Jupyter execution cell and run it:

```
from matplotlib.patches import Rectangle, Circle, Polygon
from matplotlib.text import Text
fig = plt.figure()
ax = fig.add_subplot(1,1,1)
options = dict(xlim=(0,20), ylim=(0,20), aspect='equal',
                xticks=[], yticks=[])
ax.update(options)
rect = Rectangle((3,3), width=7, height=7,
                color='orange', ec='black')
circle = Circle((13,6.5), radius=5,
                color='green', ec='darkgreen',
                alpha=0.7)
vertices = np.array([[1, 11],[10,18],
                    [15,5],[2,19]])
poly = Polygon(vertices, color='blue', lw=4,
              closed=False, fill=None)
for patch in [rect, circle, poly]:
    ax.add_patch(patch)
text = Text(12, 14, "Hello,\nWorld!", color='red',
            family='sans-serif', fontsize=26)
```

```
ax.add_artist(text)
None
```

Running this code will produce output similar to the following screenshot:

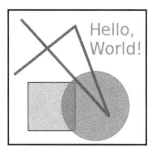

How it works...

The code starts by importing the `Artist` subclasses that we will use in the example, which include `Rectangle`, `Circle`, `Polygon`, and `Text`.

We then create a figure and add a single subplot to the figure using the `add_subplot()` method. This method returns an object of type `Axes`, which we store in the `ax` variable.

The `Axes` class is the workhorse of the Matplotlib library. Although it is possible to generate plots using lower-level calls, using the `Axes` object is sufficient for the vast majority of applications.

Next, we set the options for the `ax` object, which specify the general appearance of the graph. This is done with the `ax.update(options)` method call, where `options` is a Python dictionary. In our example, the following options are used:

- `xlim=(0,20)` and `ylim=(0,20)` set the limits for the x and y coordinates, respectively,
- `aspect='equal'` sets the scale in the vertical and horizontal directions to be the same. This prevents distortions when drawing circles and rectangles.
- `xticks=[]` and `yticks=[]` remove the tick marks from the x and y axes.

The `Axes` object allows for many other configuration options, which can be found in the API documentation at `https://matplotlib.org/api/axes_api.html`.

After configuring the `Axes` object, we create three objects for a rectangle, a circle, and a polygon. These are `Artist` objects in the `Patch` subclass, which represent a primitive graphical element that has a face color. Each object has different configuration options, which can be found in the Patch API specification, available at `https://matplotlib.org/devdocs/api/patches_api.html`.

After the patches are created, they are added to the `ax` object, by executing the following code:

```
for patch in [rect, circle, poly]:
    ax.add_patch(patch)
```

Each `Axes` object keeps several lists, containing the various graphical components of the plot. Each list corresponds to an `add_` method, that is used to add a component of the particular type. In the preceding code, the `add_patch()` method is used to add the `rect`, `circle`, and `poly` objects.

 The lists maintained by the `Axes` object should not be manipulated directly. Use the several `add_` methods to add plot elements and the corresponding `remove()` method to remove them.

The next steps in the preceding code add a `Text` object to the `Axes`. We first create the `Text` object, which contains the `'Hello\nWorld!'` multiline string and set the color, font family and font size for the text. The `Text` object is then added to the graph with a call to the `add_artist()` method.

The `Artist` interface is the most powerful and flexible method to create graphs in Matplotlib. Programmers that need to define a new graph type will probably need to work with `Artist` objects. The price paid for this power is increased complexity. Readers interested in going deeper in this topic can get started with the `Artist` tutorial, available at `https://matplotlib.org/users/artists.html`.

Creating a map with cartopy

There is a multitude of libraries based on Matplotlib dedicated to the creation of specialized graphs. In this recipe, we will discuss cartopy, which is a Python package for creating maps and handling the display of geographical information.

Getting ready

The first step is to install cartopy on your system. Since cartopy assumes a large number of dependencies, this recipe assumes that you have the Anaconda distribution installed. This being the case, simply run the following statement from the command line to install cartopy:

```
conda install -c conda-forge cartopy
```

After the installation finishes, start Jupyter and run the following three commands in an execution cell:

```
%matplotlib inline
import numpy as np
import matplotlib.pyplot as plt
```

How to do it...

Enter the following in a Jupyter code cell and run it:

```
import cartopy.crs as ccrs
fig = plt.figure(figsize=(10,10))
ax = fig.add_subplot(1, 1, 1,
     projection=ccrs.Mollweide())
ax.coastlines()
ax.stock_img()
rj_lat, rj_lon = -22.91, -43.18
bj_lat, bj_lon = 39.90, 116.41
plt.plot([rj_lon, bj_lon], [rj_lat, bj_lat],
         color='brown', lw=2, marker='o', ms=10,
         transform=ccrs.Geodetic())
plt.text(rj_lon, rj_lat-8, 'Rio de Janeiro',
         fontsize=14,
         transform=ccrs.Geodetic())
plt.text(bj_lon-7, bj_lat-9, 'Beijing',
         fontsize=14, horizontalalignment='right',
         transform=ccrs.Geodetic())
None
```

When this code is executed, we get the following output:

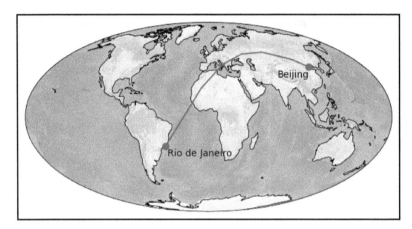

How it works...

In the first line of code, we import the `cartopy.crs` module with the `ccrs` alias. We then create a `Figure` object and an `Axes` object with the following call:

```
ax = fig.add_subplot(1, 1, 1,
    projection=ccrs.Mollweide())
```

Notice that we need to specify the `projection` option in the `add_subplot()` method call. Cartopy makes available several different projections for drawing global maps.

Next, we add to the figure the continent coastlines and a stock image of the globe with the following commands:

```
ax.coastlines()
ax.stock_img()
```

 The first time you run these commands, cartopy downloads the data it needs to produce the map.

The next step is to add data to the map. In a more realistic example, we would be using a dataset that includes some geographical information. In this example, we simply mark two cities in the globe and draw the shortest path on the Earth's surface connecting these two cities.

We start by defining the longitude and latitude of the cities. Then, we use the standard Matplotlib `plot()` and `text()` functions to add the graphical elements to the map. Notice the use of `transform=ccrs.Geodetic()` to map the coordinates to the correct locations on the map.

As demonstrated in this example, cartopy has a simple interface to generate maps and add geographical information. The full documentation of the API can be found at `http://scitools.org.uk/cartopy/`.

4

Data Wrangling with pandas

In this chapter, we will see the following recipes:

- Creating Series objects
- Creating DataFrame objects
- Inserting and deleting columns to a DataFrame
- Inserting and deleting rows to a DataFrame
- Selecting items by row indexes and column labels
- Selecting items by integer location
- Selecting items using mixed indexing
- Accessing, selecting, and modifying data
- Selecting rows using Boolean selection
- Reading and storing data in different formats
- Data displays employing different kinds of visual representation
- How to apply numerical functions and operations to Series and DataFrame objects
- Computing statistical functions on Series and DataFrame objects
- How to sort data in Series and DataFrame objects
- Performing merging, joins, concatenation, and grouping

Creating Series objects

In the following section, we will see how to create a series of objects.

Getting ready

Before trying this recipe, import the numpy and pandas modules with the following code:

```
import numpy as np
import pandas as pd
```

How to do it...

Run the following code in a code cell; this will create three objects of Series type:

```
series1 = pd.Series(['Alice', 'Bob', 'Joe', 'Mary'])
series2 = pd.Series(['Honda', 'Honda', 'Honda', 'Toyota', 'Ford'],
                    index=['CR-V', 'Civic', 'Accord', 'Highlander',
'F-150'])
series3 = pd.Series({'a':102, 'b':np.NaN, 'c':303}, dtype=np.float64)
```

After the objects are created, their contents can be checked by entering the variable that represents each object in a code cell by itself.

How it works...

The simplest constructor for a Series object takes as argument a Python iterable, as shown in the first example, reproduced as follows:

```
series1 = pd.Series(['Alice', 'Bob', 'Joe', 'Mary'])
series1
```

The iterable must be one-dimensional, such as a plain list, a NumPy array, or a file object with a stream of data. The preceding code will create the following Series object:

```
0    Alice
1      Bob
2      Joe
3     Mary
dtype: object
```

The first column in the display is the index, which is a pandas object of type `Index`. The default index uses integers, starting at 0. Also notice that a `Series` object has a specific data type, represented by a NumPy `dtype` object. In this example, the data in the `Series` object consists of strings, so the `dtype` is `object`.

In the second example, shown as follows, we specify the index for the series explicitly:

```
series2 = pd.Series(['Honda', 'Honda', 'Honda', 'Toyota', 'Ford'],
                    index=['CR-V', 'Civic', 'Accord', 'Highlander',
'F-150'])
series2
```

The index is set by using the index option in the constructor, and accepts a one-dimensional Python iterable. The data created with this command is displayed as follows:

```
CR-V            Honda
Civic           Honda
Accord          Honda
Highlander     Toyota
F-150            Ford
dtype: object
```

In the final example, we use a dictionary to specify both the index and the data, as follows:

```
series3 = pd.Series({'a':102, 'b':np.NaN, 'c':303}, dtype=np.float64)
series3
```

The keys of the dictionary become the index, and the values become the corresponding data. The `Series` object constructed by this code is displayed as follows:

```
a     102.0
b       NaN
c     303.0
dtype: float64
```

There are two other features of notice demonstrated in this example:

- The NumPy value `NaN`, which means not a number, is used to represent missing numerical data
- We specify the data type explicitly with the `dtype` option in the constructor

Creating DataFrame objects

In this section, we will see how to create DataFrame objects.

Getting ready

Before trying this recipe, import the numpy and pandas modules with the following code:

```
import numpy as np
import pandas as pd
```

How to do it...

Enter the following commands in a code cell:

```
df1 = pd.DataFrame([['Pen', 24, 2.39],
                    ['Eraser', 32, 1.29],
                    ['Sharpener', 12, 10.39],
                    ['Pencil', 42, 0.59 ]],
                index=[100024, 201024, 202034, 101122],
                columns = ['item', 'inventory', 'unit_price'])
df2 = pd.DataFrame({'item': ('Pen', 'Eraser', 'Sharpener', 'Pencil'),
                'inventory': (24, 32, 12, 42),
                'unit_price': (2.39, 1.29, 10.39, 0.59)},
                index=[100024, 201024, 202034, 101122])
```

This code shows how to generate two DataFrame objects containing essentially the same data, that could represent the inventory of an office supplies store.

How it works...

In the first preceding example, the DataFrame df1 is constructed with the following inputs:

- A list of lists, which is interpreted as a two-dimensional table. Each list represents a row of the table.
- The index option, which is a list representing the indexes for each row of the table.
- The column option, which is a list representing the column names of the table.

In the second example, the DataFrame df2 is constructed with the following inputs:

- A dictionary. Each item of the dictionary corresponds to a column of the table. The keys of the dictionary give the column names.
- The index option, which is a list representing the indexes of each row of the table.

The data contained in the two tables is essentially the same. However, notice that, since dictionaries are not ordered data structures, the order of the columns in df2 is arbitrary.

Inserting and deleting columns to a DataFrame

In this section, we will see how to insert and delete columns to a data frame.

Getting ready

Run the following code to load NumPy, and pandas, and to initialize a DataFrame object:

```
import numpy as np
import pandas as pd
df1 = pd.DataFrame([['Pen', 24, 2.39],
                    ['Eraser', 32, 1.29],
                    ['Sharpener', 12, 10.39],
                    ['Pencil', 42, 0.59 ]],
                  index=[100024, 201024, 202034, 101122],
                  columns = ['item', 'inventory', 'unit_price'])
```

How to do it...

To insert a column, run the following cell:

```
df['discount'] = [0.75, 0.50, 0.75, 0.75]
```

To delete a column, use the following code:

```
del df['discount']
```

How it works...

We can think of a DataFrame as a dictionary, where the dictionary keys are the column names. So, to insert a column, we can use the following code:

```
df['column_name'] = <iterable>
```

The <iterable> can be for example, a list, a NumPy array, or a pandas Series object. Analogously, to delete a column we can use the Python del operator, as follows:

```
del df['column_name']
```

Inserting and deleting rows to a DataFrame

In the following section, we will see how to insert and delete rows from a DataFrame.

Getting ready

Run the following code to load NumPy, and pandas, and to initialize a DataFrame object:

```
import numpy as np
import pandas as pd
df1 = pd.DataFrame([['Pen', 24, 2.39],
                    ['Eraser', 32, 1.29],
                    ['Sharpener', 12, 10.39],
                    ['Pencil', 42, 0.59 ]],
                   index=[100024, 201024, 202034, 101122],
                   columns = ['item', 'inventory', 'unit_price'])
```

How to do it...

To insert rows at the end of the DataFrame, run the code in the following cell:

```
df2 = pd.DataFrame([['Notepad', 12, 4.25],
                    ['Binder', 8, 5.68]],
                   index=[230015, 211040],
                   columns = ['item', 'inventory', 'unit_price']
df3 = df1.append(df2)
```

To delete rows from a DataFrame, use the following code:

```
df4 = df3.drop([230015, 211040])
```

How it works...

The `append()` method appends a `DataFrame` at the bottom of another `DataFrame`. The method does not alter the input data. Instead, it returns a new `DataFrame` with the concatenated data.

Analogously, the `drop()` method accepts as input a list of indexes and drops the corresponding rows in a `DataFrame`. Again, the operation is not in-place and a new `DataFrame` with the rows dropped is returned.

Selecting items by row indexes and column labels

In the following section, we will see how to select items by indexes and column labels.

Getting ready

Run the following code to load NumPy, and pandas, and to initialize a `DataFrame` object:

```python
import numpy as np
import pandas as pd
df1 = pd.DataFrame([['Pen', 24, 2.39],
                    ['Eraser', 32, 1.29],
                    ['Sharpener', 12, 10.39],
                    ['Pencil', 42, 0.59 ]],
                   index=[100024, 201024, 202034, 101122],
                   columns = ['item', 'inventory', 'unit_price'])
```

How to do it...

Run the following code, which demonstrates different ways to select data from a `DataFrame`:

```python
col = df['item']
cols = df[['item', 'inventory']]
rows = df[2:4]
row_data = df.loc[201024]
df1 = df.loc[201024:,['item','unit_price']]
```

The data selected in each of the commands is displayed as follows:

```
col:
```

```
    100024           Pen
    201024        Eraser
    202034     Sharpener
    101122        Pencil
```

```
cols:
```

```
                item   inventory
    100024       Pen          24
    201024    Eraser          32
    202034 Sharpener          12
    101122    Pencil          42
```

```
rows:
```

```
                item   inventory   unit_price
    202034 Sharpener          12        10.39
    101122    Pencil          42         0.59
```

```
row_data:
```

```
    item              Eraser
    inventory             32
    unit_price          1.29
    Name: 201024, dtype: object
```

```
df1:
```

```
                item   unit_price
    201024    Eraser         1.29
    202034 Sharpener        10.39
    101122    Pencil         0.59
```

How it works...

The `col = df['item']` line of code selects the column index from the `DataFrame` and returns in the `col` variable a `Series` object containing the specified column. Analogously, the `cols = df[['item', 'inventory']]` line selects two columns. Notice that, in this case, the object returned is a `DataFrame`.

In the next example, we use `rows = df[2:4]` to select a range of rows from the `DataFrame`. Notice that this uses the standard convention for ranges in Python and rows two and three are returned.

The next two examples show the use of the `loc` indexing method.

With the `row_data = df.loc[201024]` line we select one row of data. The returned object is a `Series`, where the indexes are the column names of the `DataFrame`.

Then, the `df1 = df.loc[201024:,['item','unit_price']]` line is used to select a sub-array of the data. The first position specifies the `201024:` slice, which represents a range starting at the `201024` index and extending to the end of the array. The second position is a list of column names, `['item','unit_price']`. The returned `DataFrame` consists in the data corresponding to the given range of rows, restricted to the specified columns.

Selecting items by integer location

In the following section, we will see how to select items by integer location.

Getting ready

Run the following code to load NumPy, and pandas, and to initialize a `DataFrame` object:

```
import numpy as np
import pandas as pd
df1 = pd.DataFrame([['Pen', 24, 2.39],
                    ['Eraser', 32, 1.29],
                    ['Sharpener', 12, 10.39],
                    ['Pencil', 42, 0.59 ]],
                  index=[100024, 201024, 202034, 101122],
                  columns = ['item', 'inventory', 'unit_price'])
```

How to do it...

Run the following code, which demonstrates different ways to select data using integer locations:

```
row_data = df.iloc[2]
col_data = df.iloc[:,1]
single_item = df.iloc[3,1]
df1 = df.iloc[0:2, 1:]
```

This code results in the following data being stored in each of the corresponding variables:

`row_data`:

```
item            Sharpener
inventory              12
unit_price          10.39
Name: 202034, dtype: object
```

`col_data`:

```
100024      24
201024      32
202034      12
101122      42
Name: inventory, dtype: int64
```

`single_item`:

```
42
```

`df1`:

```
                item  unit_price
201024         Eraser        1.29
202034      Sharpener       10.39
101122         Pencil        0.59
```

How it works...

The `iloc` indexing method from the `DataFrame` object works in the same way as numerical indexing works in two-dimensional NumPy arrays. The code illustrates several possibilities, explained in detail as follows:

- `df.iloc[2]` selects the row with index 2. As usual, indexing starts at 0, so the third row in the `DataFrame` is returned. It returns a `Series` objects that has the column names as indexes.
- `df.iloc[:,1]` selects the column with index 1, which in our example is labeled inventory. It returns a `Series` object with indexes identical to the row indexes in the original `DataFrame`.
- `df.iloc[3,1]` returns a single item from the `DataFrame`, the item at row three and column one.
- `df.iloc[0:2, 1:]` selects a subset from the `DataFrame` using ranges. The meaning of the ranges is the same as the case of NumPy arrays.
- `df.iloc[[1,2],[0,2]]` selects a subset from the `DataFrame` using lists as indexes. This particular example selects the elements where the rows at positions `[1,2]` intersect the columns at positions `[0,2]`.

Selecting items using mixed indexing

In the next section, we will see how to select items using mixed indexing.

Getting ready

Run the following code to load NumPy, and pandas, and to initialize a `DataFrame` object:

```
import numpy as np
import pandas as pd
df1 = pd.DataFrame([['Pen', 24, 2.39],
                    ['Eraser', 32, 1.29],
                    ['Sharpener', 12, 10.39],
                    ['Pencil', 42, 0.59 ]],
                   index=[100024, 201024, 202034, 101122],
                   columns = ['item', 'inventory', 'unit_price'])
```

How to do it...

Run the following commands, which provide examples of the `ix` indexing method:

```
single_item = df.ix[100024, 1]
df1 = df.ix[[201024,202034], :2]
```

The data returned by these calls is the following:

`single_item`:

    24

`df1`:

```
              item  inventory
201024      Eraser         32
202034   Sharpener         12
```

How it works...

The `ix` indexing method allows mixing integer locations with references by labels.

In the first example, `df.ix[100024, 1]`, the row is indexed by the `DataFrame` index and the column is referred to by position. Notice that, if the `DataFrame` index is an integer, `ix` interprets the location as a reference to the index. In this case, to index rows by position it is necessary to use the `iloc` method.

In the second example, `df.ix[[201024,202034], :2]`, a subset of the table is selected. The rows are specified as a list of indexes and the columns are specified as a slice given by a numerical range.

Accessing, selecting, and modifying data

In the following section, we will see some basic and advanced methods for indexing, editing, inserting, and deleting data.

Getting ready

A `DataFrame` consists of both rows and columns and has constructs to select data from specific rows and columns. These selections use the same operators as a `Series`, including `[]`, `.loc[]`, and `.iloc[]`.

Because of the multiple dimensions, the process by which these are applied differs slightly. We will examine these by first learning to select columns, then rows, a combination of rows and columns in a single statement, and also by using Boolean selections.

Additionally, pandas provides a construct for selecting a single scalar value at a specific row and column that we will investigate. This technique is important and exists because it is a very high-performance means of accessing these values.

How to do it...

Selecting the data in specific columns of a `DataFrame` is performed by using the `[]` operator. This differs from a `Series`, where `[]` specifies rows. The `[]` operator can be passed either a single object or a list of objects representing the columns to retrieve.

The following retrieves the column with the name `Sector`:

```
In [21]:  # retrieve the Sector column
          sp500['Sector'].head()

Out[21]:  Symbol
          MMM                    Industrials
          ABT                    Health Care
          ABBV                   Health Care
          ACN         Information Technology
          ACE                     Financials
          Name: Sector, dtype: object
```

When a single column is retrieved from a `DataFrame`, the result is a `Series`:

```
In [22]:  type(sp500['Sector'])

Out[22]:  pandas.core.series.Series
```

Multiple columns can be retrieved by specifying a list of column names:

```
In [23]:  # retrieve the Price and Book Value columns
          sp500[['Price', 'Book Value']].head()

Out[23]:          Price  Book Value
          Symbol
          MMM     141.14      26.668
          ABT      39.60      15.573
          ABBV     53.95       2.954
          ACN      79.79       8.326
          ACE     102.91      86.897
```

Since this has multiple columns, the result is a `DataFrame` instead of a `Series`:

```
In [24]:  # show that this is a DataFrame
          type(sp500[['Price', 'Book Value']])

Out[24]:  pandas.core.frame.DataFrame
```

Columns can also be retrieved by attribute access. A `DataFrame` can have properties added that represent the names of each column, as long as the name does not contain spaces. The following retrieves the `Price` column in this manner:

```
In [25]:  # attribute access of column by name
          sp500.Price

Out[25]:  Symbol
          MMM     141.14
          ABT      39.60
          ABBV     53.95
          ACN      79.79
          ACE     102.91
                    ...
          YHOO     35.02
          YUM      74.77
          ZMH     101.84
          ZION     28.43
          ZTS      30.53
          Name: Price, Length: 500, dtype: float64
```

Note that this will not work for the `Book  Value` column, as the name has a space.

How it works...

Selecting rows of a DataFrame

Rows can be retrieved via an index label value using `.loc[]`:

```
In [26]:   # get row with label MMM
           # returned as a Series
           sp500.loc['MMM']

Out[26]:   Sector             Industrials
           Price                   141.14
           Book Value              26.668
           Name: MMM, dtype: object
```

Furthermore, multiple rows can be retrieved using a list of labels:

```
In [27]:   # rows with label MMM and MSFT
           # this is a DataFrame result
           sp500.loc[['MMM', 'MSFT']]

Out[27]:                        Sector    Price   Book Value
           Symbol
           MMM                  Industrials   141.14      26.668
           MSFT    Information Technology    40.12      10.584
```

Rows can be retrieved by location using `.iloc[]`:

```
In [28]:   # get rows in location 0 and 2
           sp500.iloc[[0, 2]]

Out[28]:              Sector    Price   Book Value
           Symbol
           MMM     Industrials   141.14      26.668
           ABBV    Health Care    53.95       2.954
```

It is possible to look up the location in the index of a specific label value and then use that value to retrieve the row by position:

```
In [29]:   # get the location of MMM and A in the index
           i1 = sp500.index.get_loc('MMM')
           i2 = sp500.index.get_loc('A')
           (i1, i2)

Out[29]:   (0, 10)
```

```
In [30]:   # and get the rows
           sp500.iloc[[i1, i2]]

Out[30]:              Sector    Price   Book Value
           Symbol
           MMM     Industrials   141.14      26.668
           A       Health Care    56.18      16.928
```

As a final note in this section, these operations are also possible using `.ix[]`. However, this has been deprecated. For more details,
see `http://pandas.pydata.org/pandas-docs/stable/indexing.html#different-choices-for-indexing`.

Selecting rows using Boolean selection

Rows can be selected by using Boolean selection. When applied to a `DataFrame`, a Boolean selection can utilize data from multiple columns.

How to do it...

Consider the following query, which identifies all stocks with a price less than `100`:

```
In [35]: # what rows have a price < 100?
         sp500.Price < 100

Out[35]: Symbol
         MMM      False
         ABT      True
         ABBV     True
         ACN      True
         ACE      False
                  ...
         YHOO     True
         YUM      True
         ZMH      False
         ZION     True
         ZTS      True
         Name: Price, Length: 500, dtype: bool
```

This result can then be applied to the `DataFrame` using the `[]` operator to return only the rows where the result was true:

```
In [36]: # now get the rows with Price < 100
         sp500[sp500.Price < 100]

Out[36]:                         Sector  Price  Book Value
         Symbol
         ABT                Health Care  39.60      15.573
         ABBV               Health Care  53.95       2.954
         ACN     Information Technology  79.79       8.326
         ADBE    Information Technology  64.30      13.262
         AES                  Utilities  13.61       5.781
         ...                        ...    ...         ...
         XYL                Industrials  38.42      12.127
         YHOO    Information Technology  35.02      12.768
         YUM     Consumer Discretionary  74.77       5.147
         ZION                Financials  28.43      30.191
         ZTS                Health Care  30.53       2.150

         [407 rows x 3 columns]
```

Multiple conditions can be put together using parentheses. The following retrieves the symbols and price for all stocks with a price between 6 and 10:

```
In [37]:  # get only the Price where Price is < 10 and > 0
          r = sp500[(sp500.Price < 10) &
                    (sp500.Price > 6)] ['Price']
          r

Out[37]:  Symbol
          HCBK    9.80
          HBAN    9.10
          SLM     8.82
          WIN     9.38
          Name: Price, dtype: float64
```

It is common to perform selection using multiple variables. The following demonstrates this by finding all rows where the Sector is Health Care and the Price is greater than or equal to 100.00:

```
In [38]:  # price > 100 and in the Health Care Sector
          r = sp500[(sp500.Sector == 'Health Care') &
                    (sp500.Price > 100.00)] [['Price', 'Sector']]
          r

Out[38]:          Price       Sector
          Symbol
          ACT     213.77  Health Care
          ALXN    162.30  Health Care
          AGN     166.92  Health Care
          AMGN    114.33  Health Care
          BCR     146.62  Health Care
          ...       ...          ...
          REGN    297.77  Health Care
          TMO     115.74  Health Care
          WAT     100.54  Health Care
          WLP     108.82  Health Care
          ZMH     101.84  Health Care

          [19 rows x 2 columns]
```

Reading and storing data in different formats

In the following section, we will see how to read and store data in different formats.

Getting ready

Let's get started with the configuration of pandas.

Configuring pandas

We start with the standard imports and options for pandas to facilitate the examples:

```
In [1]:  # import numpy and pandas
         import numpy as np
         import pandas as pd

         # used for dates
         import datetime
         from datetime import datetime, date

         # Set some pandas options controlling output format
         pd.set_option('display.notebook_repr_html', False)
         pd.set_option('display.max_columns', 8)
         pd.set_option('display.max_rows', 10)
         pd.set_option('display.width', 90)

         # bring in matplotlib for graphics
         import matplotlib.pyplot as plt
         %matplotlib inline
```

How to do it...

We will see different ways on how to report data and how use it in different applications:

Working with CSV, text/tabular, and format data

CSV formatted data is likely to be one of the most common forms of data you may use in pandas. Many web-based services provide data in a CSV format, as do many information systems within an enterprise. It is an easy format to use and is commonly used as an export format for spreadsheet applications such as Excel.

A CSV is a file consisting of multiple lines of text-based data, with values separated by commas. It can be thought of as a table of data similar to a single sheet in a spreadsheet program. Each row of data is in its own line in the file, and each column for each row is stored in text format, with a comma separating the data in each column.

 For more details on the specifics of CSV files, feel free to visit `http://en.wikipedia.org/wiki/Comma-separated_values`.

As CSV is so common and easily understood, we will spend most of the time describing how to read and write pandas data in this format. Lessons learned from CSV methods will apply to the other formats as well and allow a little more expediency when covering these other formats.

How it works...

We will start by reading a simple CSV file, `msft.csv`. This file is a snapshot of stock values for the MSFT ticker

Reading a CSV file into a DataFrame

The data in `msft.csv` is perfect to read into a `DataFrame`. All its data is complete and has column names in the first row. All that we need to do to read this data into a `DataFrame` is to use the pandas `pd.read_csv()` function:

```
In [3]:  # read in msft.csv into a DataFrame
         msft = pd.read_csv("data/msft.csv")
         msft[:5]

Out[3]:         Date   Open   High    Low  Close    Volume
         0  7/21/2014  83.46  83.53  81.81  81.93   2359300
         1  7/18/2014  83.30  83.40  82.52  83.35   4020800
         2  7/17/2014  84.35  84.63  83.33  83.63   1974000
         3  7/16/2014  83.77  84.91  83.66  84.91   1755600
         4  7/15/2014  84.30  84.38  83.20  83.58   1874700
```

Wow, that was easy! pandas has realized that the first line of the file contains the names of the columns and bulk read in the data to `DataFrame`.

Specifying the index column when reading a CSV file

In the previous example, the index is numerical, starting from 0, instead of being ordered by `Date`. This is because pandas does not assume that any specific column in the file should be used as the index. To help this situation, you can specify which column(s) should be the index in the call to `read_csv()` using the `index_col` parameter by assigning it the zero-based position of the column to be used as the index.

The following reads the data and tells pandas to use the column at position 0 in the file as the index (the `Date` column):

```
In [4]:  # use column 0 as the index
         msft = pd.read_csv("data/msft.csv", index_col=0)
         msft[:5]

Out[4]:            Open   High    Low  Close    Volume
         Date
         7/21/2014  83.46  83.53  81.81  81.93   2359300
         7/18/2014  83.30  83.40  82.52  83.35   4020800
         7/17/2014  84.35  84.63  83.33  83.63   1974000
         7/16/2014  83.77  84.91  83.66  84.91   1755600
         7/15/2014  84.30  84.38  83.20  83.58   1874700
```

Reading and writing data in Excel format

pandas supports reading data in Excel 2003 and newer formats, using the pd.read_excel() function or via the ExcelFile class. Internally, both techniques use either the XLRD or OpenPyXL packages, so you will need to ensure that one of them is installed in your Python environment.

For demonstration, a stocks.xlsx file is provided with the sample data. If you open it in Excel, you will see something similar to what is shown in the following screenshot:

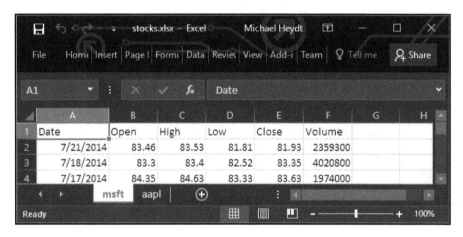

The workbook contains two sheets, **msft** and **aapl**, which hold the stock data for each respective stock.

The following then reads the stocks.xlsx file into a DataFrame:

```
In [19]:  # read excel file
          # only reads first sheet (msft in this case)
          df = pd.read_excel("data/stocks.xlsx")
          df[:5]

Out[19]:        Date   Open   High    Low  Close   Volume
          0 2014-07-21  83.46  83.53  81.81  81.93  2359300
          1 2014-07-18  83.30  83.40  82.52  83.35  4020800
          2 2014-07-17  84.35  84.63  83.33  83.63  1974000
          3 2014-07-16  83.77  84.91  83.66  84.91  1755600
          4 2014-07-15  84.30  84.38  83.20  83.58  1874700
```

This has read only content from the first worksheet in the Excel file (the **msft** worksheet), and has used the contents of the first row as column names. To read the other worksheet, you can pass the name of the worksheet using the `sheetname` parameter:

```
In [20]:  # read from the aapl worksheet
          aapl = pd.read_excel("data/stocks.xlsx", sheetname='aapl')
          aapl[:5]

Out[20]:          Date   Open   High    Low  Close     Volume
          0 2014-07-21  94.99  95.00  93.72  93.94  38887700
          1 2014-07-18  93.62  94.74  93.02  94.43  49898600
          2 2014-07-17  95.03  95.28  92.57  93.09  57152000
          3 2014-07-16  96.97  97.10  94.74  94.78  53396300
          4 2014-07-15  96.80  96.85  95.03  95.32  45477900
```

Like with `pd.read_csv()`, many assumptions are made about column names, data types, and indexes. All the options we covered for `pd.read_csv()` to specify this information, also apply to the `pd.read_excel()` function.

Excel files can be written using the `.to_excel()` method of `DataFrame`. Writing to the XLS format requires the inclusion of the `XLWT` package; so, make sure it is loaded in your Python environment before trying.

The following writes the data we just acquired to `stocks2.xls`. The default is to store `DataFrame` in the **Sheet1** worksheet:

```
In [21]:  # save to an .XLS file, in worksheet 'Sheet1'
          df.to_excel("data/stocks2.xls")
```

Opening this in Excel shows you the following:

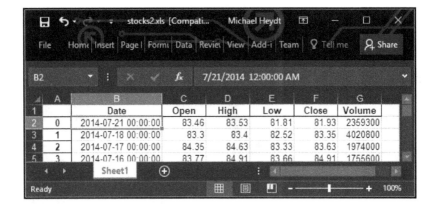

You can specify the name of the worksheet using the `sheet_name` parameter:

```
In [22]:  # write making the worksheet name MSFT
          df.to_excel("data/stocks_msft.xls", sheet_name='MSFT')
```

In Excel, we can see that the sheet has been named **MSFT**:

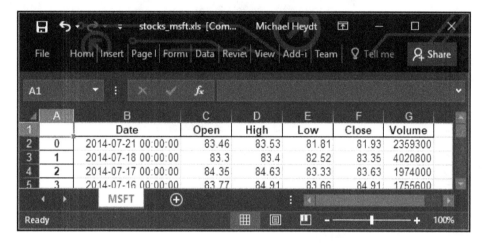

To write more than one `DataFrame` to a single Excel file using each `DataFrameobject` on a separate worksheet, use the `ExcelWriter` object along with the `with` keyword. `ExcelWriter` is part of pandas, but you will need to make sure it is imported, as it is not in the top-level namespace of pandas. The following writes two `DataFrame` objects to two different worksheets in one Excel file:

```
In [23]:  # write multiple sheets
          # requires use of the ExcelWriter class
          from pandas import ExcelWriter
          with ExcelWriter("data/all_stocks.xls") as writer:
              aapl.to_excel(writer, sheet_name='AAPL')
              df.to_excel(writer, sheet_name='MSFT')
```

We can see that there are two worksheets in the Excel file:

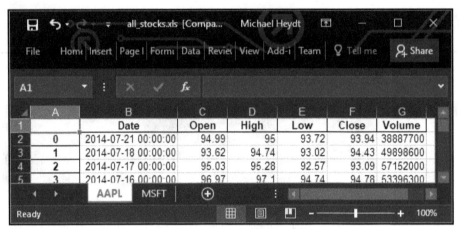

Writing to XLSX files uses the same function, but specifies `.xlsx` as the file extension:

```
In [24]: # write to xlsx
         df.to_excel("data/msft2.xlsx")
```

Reading and writing JSON files

pandas can read and write data stored in the **JavaScript Object Notation (JSON)** format. This is one of my favorites, due to its ability to be used across platforms and with many programming languages.

To demonstrate saving as JSON, we will first save the Excel data we just read into a JSON file and examine the contents:

```
In [25]:  # wirite the excel data to a JSON file
          df[:5].to_json("data/stocks.json")
          !cat data/stocks.json # osx or Linux
          #type data/stocks.json # windows

          {"Date":{"0":1405900800000,"1":1405641600000,"2":140555520000
          0,"3":1405468800000,"4":1405382400000},"Open":{"0":83.46,"1":83.
          3,"2":84.35,"3":83.77,"4":84.3},"High":{"0":83.53,"1":83.4,"2":8
          4.63,"3":84.91,"4":84.38},"Low":{"0":81.81,"1":82.52,"2":83.3
          3,"3":83.66,"4":83.2},"Close":{"0":81.93,"1":83.35,"2":83.6
          3,"3":84.91,"4":83.58},"Volume":{"0":2359300,"1":4020800,"2":197
          4000,"3":1755600,"4":1874700},"Adj Close":{"0":81.93,"1":83.3
          5,"2":83.63,"3":84.91,"4":83.58}}
```

JSON-based data can be read with the `pd.read_json()` function:

```
In [26]:  # read data in from JSON
          df_from_json = pd.read_json("data/stocks.json")
          df_from_json[:5]

Out[26]:    Adj Close  Close        Date   High    Low   Open   Volume
          0     81.93  81.93  2014-07-21  83.53  81.81  83.46  2359300
          1     83.35  83.35  2014-07-18  83.40  82.52  83.30  4020800
          2     83.63  83.63  2014-07-17  84.63  83.33  84.35  1974000
          3     84.91  84.91  2014-07-16  84.91  83.66  83.77  1755600
          4     83.58  83.58  2014-07-15  84.38  83.20  84.30  1874700
```

Note the two slight differences here, caused by the reading/writing of data from JSON. First, the columns have been reordered alphabetically. Second, the index for `DataFrame`, although containing content, is sorted as a string. These issues can be fixed easily, but they will not be covered here for brevity.

Reading HTML data from the web

pandas has support for reading data from HTML files (or HTML from URLs). Underneath the covers, pandas makes use of the `LXML`, `Html5Lib`, and `BeautifulSoup4` packages. These packages provide some impressive capabilities for reading and writing HTML tables.

Your default installation of Anaconda may not include these packages. If you get errors using this function, install the appropriate library based on the error, using Anaconda Navigator:

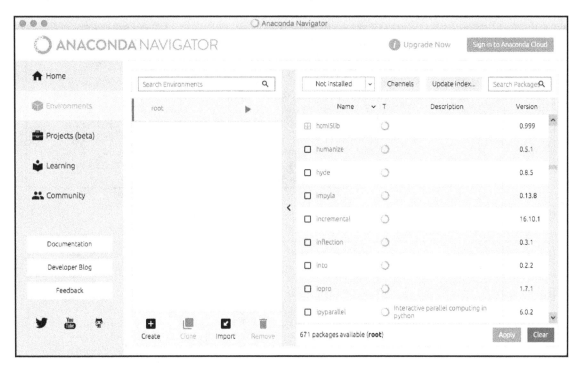

Otherwise, you can use `pip`:

```
Michaels-iMac-2:~ michaelheydt$ /Users/michaelheydt/.anaconda/navigator/a.tool ; exit;
[(/Users/michaelheydt/anaconda) bash-3.2$ pip install html5lib
Collecting html5lib
  Downloading html5lib-0.999999999-py2.py3-none-any.whl (112kB)
    100% |████████████████████████████████| 122kB 126kB/s
Requirement already satisfied: six in ./anaconda/lib/python3.6/site-packages (from html5
lib)
Collecting webencodings (from html5lib)
  Downloading webencodings-0.5.1-py2.py3-none-any.whl
Requirement already satisfied: setuptools>=18.5 in ./anaconda/lib/python3.6/site-package
s/setuptools-27.2.0-py3.6.egg (from html5lib)
Installing collected packages: webencodings, html5lib
Successfully installed html5lib-0.999999999 webencodings-0.5.1
(/Users/michaelheydt/anaconda) bash-3.2$ █
```

The `pd.read_html()` function will read HTML from a file (or URL) and parse all HTML tables found in the content into one or more pandas `DataFrame` objects. The function always returns a list of `DataFrame` objects (actually, zero or more, depending on the number of tables found in the HTML).

To demonstrate, we will read table data from the FDIC failed bank list, located at `https://www.fdic.gov/bank/individual/failed/banklist.html`. Viewing the page, you can see there is a list of quite a few failed banks.

This data is actually very simple to read with pandas and its `pd.read_html()` function:

```
In [27]:  # the URL to read
          url = "http://www.fdic.gov/bank/individual/failed/banklist.html"
          # read it
          banks = pd.read_html(url)
```

```
In [28]:  # examine a subset of the first table read
          banks[0][0:5].iloc[:,0:2]
Out[28]:                                      Bank Name              City
          0                         Fayette County Bank        Saint Elmo
          1  Guaranty Bank, (d/b/a BestBank in Georgia & Mi...   Milwaukee
          2                              First NBC Bank       New Orleans
          3                              Proficio Bank  Cottonwood Heights
          4            Seaway Bank and Trust Company           Chicago
```

Again, that was almost too easy!

A `DataFrame` can be written to an HTML file with the `.to_html()` method. This method creates a file containing only the `<table>` tag for the data (not the entire HTML document). The following writes the stock data we read earlier to an HTML file:

```
In [28]:   # read the stock data
           df = pd.read_excel("data/stocks.xlsx")
           # write the first two rows to HTML
           df.head(2).to_html("data/stocks.html")
           # check the lines of the output
           df_from_html = pd.read_html("data/stocks.html")
           df_from_html

Out[28]:   [   Unnamed: 0        Date   Open   High    Low  Close   Volume
           0            0  2014-07-21  83.46  83.53  81.81  81.93  2359300
           1            1  2014-07-18  83.30  83.40  82.52  83.35  4020800]
```

Viewing this in the browser looks like what is shown in the following screenshot:

This is useful, as you can use pandas to write HTML fragments to be included in websites, update them when needed, and thereby have the new data available to the site statically instead of through a more complicated data query or service call.

Accessing CSV data on the web

It is quite common to read data off the web and from the internet. pandas makes it easy to read data from the web. All the pandas functions that we have examined can also be given an HTTP URL, FTP address, or S3 address, instead of a local file path, and all of them work just the same as they work with a local file.

The following demonstrates how easy it is to directly make HTTP requests using the existing pd.read_csv() function:

```
In [32]: # Reading data from a web site
         database_countries = pd.read_csv("https://raw.githubusercontent.com/cs109/2014_data/master/countries.csv", sep=",")
         database_countries

Out[32]:          Country        Region
         0        Algeria        AFRICA
         1        Angola         AFRICA
         2        Benin          AFRICA
         3        Botswana       AFRICA
         4        Burkina        AFRICA
         ..       ...            ...
         189      Paraguay       SOUTH AMERICA
         190      Peru           SOUTH AMERICA
         191      Suriname       SOUTH AMERICA
         192      Uruguay        SOUTH AMERICA
         193      Venezuela      SOUTH AMERICA

         [194 rows x 2 columns]
```

Reading and writing from/to SQL databases

pandas can read data from any SQL database that supports Python data adapters that respect the Python DB-API. Reading is performed using the pandas.io.sql.read_sql() function, and writing to SQL databases is done using the .to_sql() method of DataFrame.

To demonstrate, the following reads the stock data from `msft.csv` and `aapl.csv`. It then makes a connection to an SQLite3 database file. If the file does not exist, it is created on the fly. It then writes the MSFT data to a table named `STOCK_DATA`. If the table does not exist, it is created as well. If it does exist, all the data is replaced with the MSFT data. Finally, it then appends the AAPL stock data to that table:

```
In [34]:  # reference SQLite
          import sqlite3

          # read in the stock data from CSV
          msft = pd.read_csv("data/msft.csv")
          msft["Symbol"]="MSFT"
          aapl = pd.read_csv("data/aapl.csv")
          aapl["Symbol"]="AAPL"

          # create connection
          connection = sqlite3.connect("data/stocks.sqlite")
          # .to_sql() will create SQL to store the DataFrame
          # in the specified table.  if_exists specifies
          # what to do if the table already exists
          msft.to_sql("STOCK_DATA", connection, if_exists="replace")
          aapl.to_sql("STOCK_DATA", connection, if_exists="append")

          # commit the SQL and close the connection
          connection.commit()
          connection.close()
```

To demonstrate that this data was created, you can open the database file with a tool such as SQLite Data Browser (available at `https://github.com/sqlitebrowser/sqlitebrowser`). The following screenshot shows you a few rows of the data in the database file:

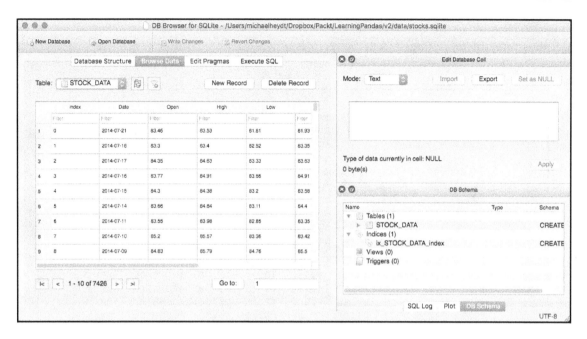

Data can be read using SQL from the database using the `pd.io.sql.read_sql()` function. The following demonstrates a query of the data from `stocks.sqlite` using SQL and reports it to the user:

```
In [35]:   # connect to the database file
           connection = sqlite3.connect("data/stocks.sqlite")

           # query all records in STOCK_DATA
           # returns a DataFrame
           # inde_col specifies which column to make the DataFrame index
           stocks = pd.io.sql.read_sql("SELECT * FROM STOCK_DATA;",
                                          connection, index_col='index')

           # close the connection
           connection.close()

           # report the head of the data retrieved
           stocks[:5]

Out[35]:             Date    Open   High    Low  Close   Volume  Symbol
           index
           0      7/21/2014  83.46  83.53  81.81  81.93  2359300   MSFT
           1      7/18/2014  83.30  83.40  82.52  83.35  4020800   MSFT
           2      7/17/2014  84.35  84.63  83.33  83.63  1974000   MSFT
           3      7/16/2014  83.77  84.91  83.66  84.91  1755600   MSFT
           4      7/15/2014  84.30  84.38  83.20  83.58  1874700   MSFT
```

It is also possible to use the WHERE clause in the SQL as well as to select columns. To demonstrate, the following selects the records where MSFT's volume is greater than 29200100:

```
In [36]:  # open the connection
          connection = sqlite3.connect("data/stocks.sqlite")
          # construct the query string
          query = "SELECT * FROM STOCK_DATA WHERE " + \
                  "Volume>29200100 AND Symbol='MSFT';"
          # execute and close connection
          items = pd.io.sql.read_sql(query, connection, index_col='index')
          connection.close()
          # report the query result
          items

Out[36]:             Date    Open   High    Low   Close     Volume  Symbol
          index
          1081    5/21/2010  42.22  42.35  40.99  42.00   33610800    MSFT
          1097    4/29/2010  46.80  46.95  44.65  45.92   47076200    MSFT
          1826    6/15/2007  89.80  92.10  89.55  92.04   30656400    MSFT
          3455    3/16/2001  47.00  47.80  46.10  45.33   40806400    MSFT
          3712    3/17/2000  49.50  50.00  48.29  50.00   50860500    MSFT
```

A final point is that most of the code in these examples was SQLite3 code. The only pandas part of these examples is the use of the .to_sql() and .read_sql() methods, as these functions take a connection object, which can be any Python DB-API-compatible data adapter; you can more or less work with any supported database data by simply creating an appropriate connection object. The code at the pandas level should remain the same for any supported database.

Data displays employing different kinds of visual representation

In the following section, we will cover different ways of displaying data employing different kinds of visual representation.

Getting ready

Let's begin our introduction with a look at the anatomy of a Matplotlib plot, in the following figure:

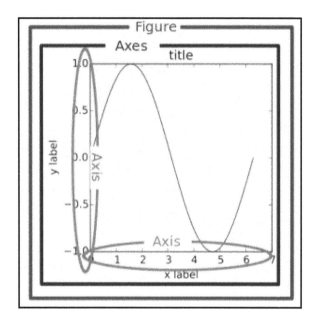

Matplotlib uses a hierarchy of objects to display all of its plotting items in the output. This hierarchy is key to understanding everything about Matplotlib. The **Figure** and **Axes** objects are the two main components of the hierarchy. The **Figure** object is at the top of the hierarchy. It is the container for everything that will be plotted. Contained within the **Figure** is one or more **Axes** object(s). The **Axes** is the primary object that you will interact with when using Matplotlib and can be more commonly thought of as the actual plotting surface. The Axes contains the x or y axis, points, lines, markers, labels, legends, and any other useful item that is plotted.

How to do it...

One of the most common data visualizations is of time-series data. Visualizing a time series in pandas is as simple as calling `.plot()` on a `DataFrame` or `Series` object that models a time series.

These are the parameters of the `Series.plot` function:

Series.plot(*kind='line', ax=None, figsize=None, use_index=True, title=None, grid=None, legend=False, style=None, logx=False, logy=False, loglog=False, xticks=None, yticks=None, xlim=None, ylim=None, rot=None, fontsize=None, colormap=None, table=False, yerr=None, xerr=None, label=None, secondary_y=False, **kwds*)

The following example demonstrates creating a time series that represents a random walk of values over time, akin to the movements in the price of a stock:

```
In [3]:  # generate a random walk time-series
         np.random.seed(seedval)
         s = pd.Series(np.random.randn(1096),
                     index=pd.date_range('2012-01-01',
                                         '2014-12-31'))
         walk_ts = s.cumsum()
         # this plots the walk - just that easy :)
         walk_ts.plot();
```

The .plot() method on pandas objects is a wrapper function around the Matplotlib library's plot() function. It makes plots of pandas data very easy to create as its implementation is coded to know how to render many visualizations based on the underlying data. It handles many of the details such as selecting series, labeling, and axis generation.

In the previous example, .plot() determined that the Series contains dates for its index and therefore the *x* axis should be formatted as dates. It also selects a default color for the data.

Plotting a Series of data gives a similar result as rendering a DataFrame ;with a single column. The following code shows this by producing the same graph with one small difference: it has added a legend to the graph. Charts will contain a legend by default when generated from a DataFrame:

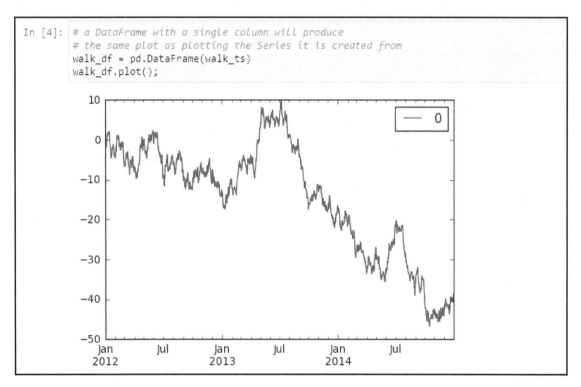

How it works...

This legend guide is an extension of the documentation available at legend()—please ensure you are familiar with contents of the documentation (`https://matplotlib.org/api/pyplot_api.html#matplotlib.pyplot.legend`) before proceeding with this guide.

xlabel() and ylabel() function parameters are listed in following tables:

```
Axes.set_xlabel(xlabel, fontdict=None, labelpad=None, **kwargs)
```

Set the label for the *x* axis.

Parameters	`xlabel`: string, x label `labelpad`: scalar, optional, default: None Spacing in points between the label and the *x* axis

```
Axes.set_ylabel(ylabel, fontdict=None, labelpad=None, **kwargs)
```

Set the label for the *y* axis:

Parameters	`ylabel`: string, y label `labelpad`: scalar, optional, default: None Spacing in points between the label and the *x* axis

This guide makes use of some common terms, which are documented here for clarity:

Legend, entry: A legend is made up of one or more legend entries. An entry is made up of exactly one key and one `label`. `legend` key. The colored/patterned marker to the left of each legend `label`. `legend` label. The text which describes the handle represented by the `key`. `legend` handle.

The original object which is used to generate an appropriate entry in the legend.

```
In [6]:  # plot the DataFrame, which will plot a line
         # for each column, with a legend
         walk_df.plot();
```

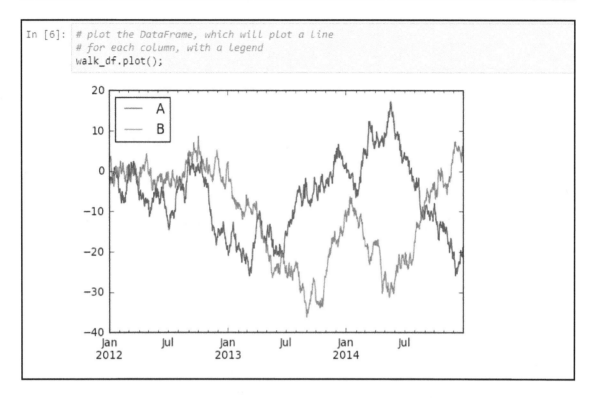

To change the text used for each data series in the legend (the default is the column name from `DataFrame`), capture the `ax` object returned from the `.plot()` method and use its `.legend()` method. This object is an `AxesSubplot` object and can be used to change various aspects of the plot before it is generated:

The location of the legend can be set using the `loc` parameter of `.legend()`. By default, pandas sets the location to `best`, which tells Matplotlib to examine the data and determine the best place it thinks to put the legend. However, you can also specify any of the following to position the legend more specifically (you can use either the string or the numeric code):

Text	Code
best	0
upper right	1
upper left	2
lower left	3

lower right	4
'right	5
center left	6
center right	7
lower center	8
upper center	9
center	10

The following example demonstrates placing the legend in the upper-center portion of the graph:

```
In [11]:  # copy the walk
          df2 = walk_df.copy()
          # add a column C which is 0 .. 1096
          df2['C'] = pd.Series(np.arange(0, len(df2)), index=df2.index)
          # instead of dates on the x-axis, use the 'C' column,
          # which will label the axis with 0..1000
          df2.plot(x='C', y=['A', 'B']);
```

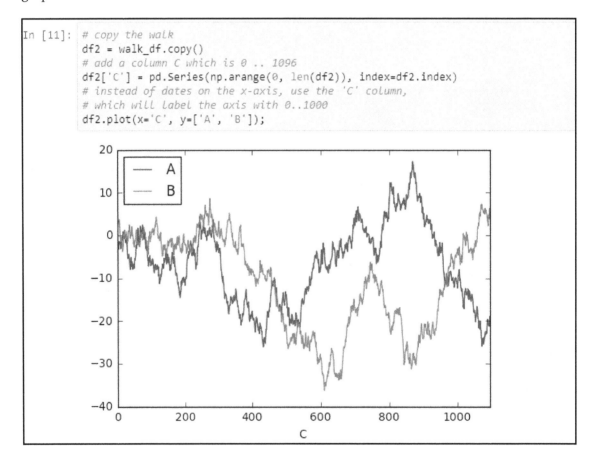

How to apply numerical functions and operations to Series and DataFrame objects

In the following section, we will see how to apply numerical functions and operations to `Series` and `DataFrame` objects.

Getting ready

When a `DataFrame` operates directly with one of the arithmetic or comparison operators, each value of each column gets the operation applied to it. Typically, when an operator is used with a `DataFrame`, the columns are either all numeric or all object (usually strings). If the `DataFrame` does not contain homogeneous data, then the operation is likely to fail. Let's see an example of this failure with the college dataset, which contains both numeric and object data types. Attempting to add 5 to each value of the `DataFrame` raises a `TypeError`, as integers cannot be added to strings:

```
college = pd.read_csv('data/college.csv')
college + 5
TypeError: Could not operate 5 with block values must be str, not int
```

To successfully use an operator with a `DataFrame`, first select homogeneous data. For this recipe, we will select all the columns that begin with `UGDS_`. These columns represent the fraction of undergraduate students by race. To get started, we import the data and use the institution name as the label for our index and then select the columns we desire with the filter method:

```
college = pd.read_csv('data/college.csv', index_col='INSTNM')
college_ugds_ = college.filter(like='UGDS_')
college_ugds_.head()
```

INSTNM	UGDS_WHITE	UGDS_BLACK	UGDS_HISP	UGDS_ASIAN	UGDS_AIAN	UGDS_NHPI	UGDS_2MOR	UGDS_NRA	UGDS_UNKN
Alabama A & M University	0.0333	0.9353	0.0055	0.0019	0.0024	0.0019	0.0000	0.0059	0.0138
University of Alabama at Birmingham	0.5922	0.2600	0.0283	0.0518	0.0022	0.0007	0.0368	0.0179	0.0100
Amridge University	0.2990	0.4192	0.0069	0.0034	0.0000	0.0000	0.0000	0.0000	0.2715
University of Alabama in Huntsville	0.6988	0.1255	0.0382	0.0376	0.0143	0.0002	0.0172	0.0332	0.0350
Alabama State University	0.0158	0.9208	0.0121	0.0019	0.0010	0.0006	0.0098	0.0243	0.0137

This recipe uses multiple operators with a `DataFrame` to round the undergraduate columns to the nearest hundredth. We will then see how this result is equivalent to the `round` method.

How to do it...

To begin our rounding adventure with operators, we will first add `.00501` to each value of `college_ugds_`:

```
college_ugds_ + .00501
```

How it works...

1. We use the plus operator, which attempts to add a scalar value to each value of each column of the `DataFrame`. As the columns are all numeric, this operation works as expected. There are some missing values in each of the columns but they stay missing after the operation.

   Mathematically, adding `.005` should be enough so that the floor division in the next step correctly rounds to the nearest whole percentage. The trouble appears because of the inexactness of floating-point numbers:

   ```
   .045 + .005
    0.049999999999999996
   ```

   There is an extra `.00001` added to each number to ensure that the floating-point representation the same first four digits as the actual value. This works because the maximum precision of all the points in the dataset is four decimal places.

2. We apply the floor division operator, `//`, to all the values in the `DataFrame`. As we are dividing by a fraction, in essence, it is multiplying each value by 100 and truncating any decimals. Parentheses are needed around the first part of the expression, as floor division has higher precedence than addition.
3. We use the division operator to return the decimal to the correct position.

4. We reproduce the previous steps with the round method. Before we can do this, we must again add an extra `.00001` to each DataFrame value for a different reason from step 1. NumPy and Python 3 round numbers that are exactly halfway between either side of the even number. This *ties to the even* (`http://bit.ly/2x3V5TU`) technique is not usually what is formally taught in schools. It does not consistently bias numbers to the higher side (`http://bit.ly/2zhsPy8`).

It is necessary here to round up so that both `DataFrame` values are equal. The `equals` method determines if all the elements and indexes between two `DataFrame` objects are exactly the same and returns a Boolean.

Computing statistical functions on Series and DataFrame objects

In the following section, we will see how to compute statistical functions on `Series` and `DataFrame` objects.

Getting ready

Descriptive statistics gives us the ability to understand numerous measures of data that describe a specific characteristic of the underlying data. Built into pandas are several classes of these descriptive statistical operations that can be applied to a `Series` or `DataFrame`.

Let's examine several facets of statistical analysis/techniques provided by pandas:

- Summary descriptive statistics
- Measuring central tendency: mean, median, and mode
- Variance and standard deviation

How to do it...

This section represents how to get some information about statistics and its applications, and how to calculate the central measurements of a process.

Retrieving summary descriptive statistics

pandas objects provide the `.describe()` method, which returns a set of summary statistics of the object's data. When applied to a `DataFrame`, `.describe()` will calculate the summary statistics for each column. The following code calculates these statistics for both stocks in `omh`:

```
In [24]:   # get summary statistics for the stock data
           omh.describe()

Out[24]:               MSFT           AAPL
           count   22.000000     22.000000
           mean    47.493182    112.411364
           std      0.933077      2.388772
           min     45.160000    106.750000
           25%     46.967500    111.660000
           50%     47.625000    112.530000
           75%     48.125000    114.087500
           max     48.840000    115.930000
```

With one quick method call, we have calculated the count, mean, standard deviation, minimum, and maximum values, and even the 25th, 50th, and 75th percentiles for both series of stock data.

`.describe()` can also be applied to a `Series`. The following code calculates summary statistics for just MSFT:

```
In [25]:   # just the stats for MSFT
           omh.MSFT.describe

Out[25]:   count    22.000000
           mean     47.493182
           std       0.933077
           min      45.160000
           25%      46.967500
           50%      47.625000
           75%      48.125000
           max      48.840000
           Name: MSFT, dtype: float64
```

Only the mean can be obtained, as follows:

```
In [26]:  # only the mean for MSFT
          omh.MSFT.describe ['mean']

Out[26]:  47.493181818181817
```

Non-numerical data will result in a slightly different set of summary statistics, returning the total number of items (count), the count of unique values (unique), most frequently occurring value (top), and the number of times it appears (freq):

```
In [27]:  # get summary stats on non-numeric data
          s = pd.Series(['a', 'a', 'b', 'c', np.NaN])
          s.describe()

Out[27]:  count      4
          unique     3
          top        a
          freq       2
          dtype: object
```

How it works...

This section represents the way we calculate the mean.

Calculating the mean

The mean, commonly referred to as the average, gives us a measurement of the **central tendency** of data. It is determined by summing up all the measurements and then dividing by the number of measurements.

The mean can be calculated using .mean(). The following code calculates the average of the prices for AAPL and MSFT:

```
In [28]:  # the mean of all the columns in omh
          omh.mean()

Out[28]:  MSFT      47.493182
          AAPL     112.411364
          dtype: float64
```

pandas has taken each column and independently calculated the mean for each. It returned the results as values in a `Series` that is indexed with the column names. The default is to apply the method on `axis=0`, applying the function to each column. This code switches the `axis` and returns the average price for all stocks on each day:

```
In [29]: # calc the mean of the values in each row
         omh.mean(axis=1)[:5]

Out[29]: 0    81.845
         1    81.545
         2    82.005
         3    82.165
         4    81.710
         dtype: float64
```

Calculating variance and standard deviation

In probability theory and statistics, standard deviation and variance give us a feel of how far some numbers are spread out from their mean. Let's briefly examine each.

Measuring variance

Variance gives us a feel for the overall amount of spread of the values from the mean. It is defined as follows:

$$S^2 = \frac{1}{N-1} \sum_{i=1}^{N} (x_i - \bar{x})^2$$

Essentially, this is stating that for each measurement, we calculate the value of the difference between the value and the mean. This can be a positive or negative value, so we square the result to make sure that negative values have cumulative effects on the result. These values are then summed up and divided by the number of measurements minus one, giving an approximation of the average value of the differences.

In pandas, the variance is calculated using the .var() method. The following code calculates the variance of the price for both stocks:

```
In [33]:  # calc the variance of the values in each column
          omh.var()

Out[33]:  MSFT    0.870632
          AAPL    5.706231
          dtype: float64
```

Finding the standard deviation

Standard deviation is a similar measurement to variance. It is determined by calculating the square root of the variance and is defined as follows:

$$S = \sqrt{\frac{1}{N-1}\sum_{i=1}^{N}(x_i - \bar{x})^2}$$

Remember that the variance squares the difference between all measurements and the mean. Because of this, the variance is not in the same units and the actual values. By using the square root of the variance, the standard deviation is in the same units as the values in the original dataset.

The standard deviation is calculated using the .std() method, as demonstrated here:

```
In [34]:  # standard deviation
          omh.std()

Out[34]:  MSFT    0.933077
          AAPL    2.388772
          dtype: float64
```

Determining correlation

Covariance can help determine whether values are related, but it does not give a sense of the degree to which the variables move together. To measure the degree to which variables move together, we need to calculate the correlation. Correlation is calculated by dividing the covariance by the product of the standard deviations of both sets of data:

$$r(x, y) = \frac{cov_{x,y}}{S_x S_y}$$

Correlation standardizes the measure of interdependence between two variables and consequently tells you how closely the two variables move. The correlation measurement, called the correlation coefficient, will always take a value between one and -1, and the interpretation of this value is as follows:

- If the correlation coefficient is 1.0, the variables have a perfect positive correlation. This means that if one variable moves by a given amount, the second moves proportionally in the same direction. A positive correlation coefficient of less than 1.0 but greater than 0.0 indicates a less-than-perfect positive correlation, with the strength of the correlation growing as the number approaches 1.0.
- If the correlation coefficient is 0.0, no relationship exists between the variables. If one variable moves, you can make no predictions about the movement of the other variable.
- If the correlation coefficient is -1.0, the variables are perfectly negatively correlated (or inversely correlated) and move opposite to each other. If one variable increases, the other variable decreases proportionally. A negative correlation coefficient greater than -1.0 but less than 0.0 indicates a less-than-perfect negative correlation, with the strength of the correlation growing as the number approaches -1.

Correlations in pandas are calculated using the .corr() method. The following code calculates the correlation of MSFT to AAPL:

```
In [36]:  # correlation of MSFT relative to AAPL
          omh.MSFT.corr(omh.AAPL)

Out[36]:  0.8641560684381171
```

This shows that the prices for MSFT and AAPL during this period demonstrate a high level of correlation. This does not mean that they are causal, with one affecting the other, but that there are likely shared influences on the values, such as being in similar markets.

How to sort data in Series and DataFrame objects

In this section, we will see how to sort data in `Series` and `DataFrame` objects.

Getting ready

One of the most basic and common operations to perform during a data analysis is to select rows containing the largest value of some column within a group. For instance, this would be like finding the highest rated film or the highest grossing film of each year by content rating. To accomplish this task, we need to sort the groups as well as the column used to rank each member of the group, and then extract the highest member of each group.

How to do it...

1. Read in the movie dataset and slim it down to just the three columns we care about, `movie_title`, `title_year`, and `imdb_score`:

    The first thing we need to do is configure pandas:

    ```
    # import numpy and pandas
    import numpy as np
    import pandas as pd

    # used for dates
    import datetime
    from datetime import datetime, date

    # Set some pandas options controlling output format
    pd.set_option('display.notebook_repr_html', False)
    pd.set_option('display.max_columns', 8)
    pd.set_option('display.max_rows', 10)
    pd.set_option('display.width', 90)

    # bring in matplotlib for graphics
    import matplotlib.pyplot as plt
    %matplotlib inline

    movie = pd.read_csv('data/movie.csv')
    movie2 = movie[['movie_title', 'title_year', 'imdb_score']]
    ```

2. Use the `sort_values` method to sort the `DataFrame` by `title_year`. The default behavior sorts from the smallest to largest. Use the ascending parameter to invert this behavior by setting it equal to `True`:

```
movie2.sort_values('title_year', ascending=False).head()
```

How it works...

1. In step 1, we slim the dataset down to concentrate on only the columns of importance. This recipe would work the same with the entire `DataFrame`.
2. Step 2 shows how to sort a `DataFrame` by a single column, which is not exactly what we wanted.
3. Step 3 sorts multiple columns at the same time. It works by first sorting all of `title_year` and then, within each distinct value of `title_year`, sorts by `imdb_score`.

The default behavior of the `drop_duplicates` method is to keep the first occurrence of each unique row, which would not drop any rows as each row is unique. However, the subset parameter alters it to only consider the column (or list of columns) given to it. In this example, only one row for each year will be returned. As we sorted by year and score in the last step, the highest scoring movie for each year is what we get.

Performing merging, joins, concatenation, and grouping

In this section, we will see how to perform merging, joins, concatenation, and grouping.

Getting ready

We start the examples in the chapter using the following imports and configuration statements:

```
In [1]:  # import numpy and pandas
         import numpy as np
         import pandas as pd

         # used for dates
         import datetime
         from datetime import datetime, date

         # Set some pandas options controlling output format
         pd.set_option('display.notebook_repr_html', False)
         pd.set_option('display.max_columns', 8)
         pd.set_option('display.max_rows', 10)
         pd.set_option('display.width', 60)

         # bring in matplotlib for graphics
         import matplotlib.pyplot as plt
         %matplotlib inline
```

How to do it...

Concatenation in pandas is the process of combining the data from two or more pandas objects into a new object. Concatenation of the `Series` objects simply results in a new `Series`, with the values copied in sequence.

The process of concatenating the `DataFrame` objects is more complex. The concatenation can be applied to either axis of the specified objects, and along that axis pandas performs relational join logic to the index labels. Then, along the opposite axis, pandas performs alignment of the labels and filling of missing values.

Because there are a number of factors to consider, we will break down the examples for concatenation into the following topics:

- Understanding the default semantics of concatenation
- Switching the axis of alignment
- Specifying the join type
- Appending data instead of concatenation
- Ignoring the index labels

How it works...

In this section, we will see the merging and how to join data using `DataFrame` method.

Merging and join data

pandas allows the merging of pandas objects with database-like join operations, using the `pd.merge()` function and the `.merge()` method of a `DataFrame` object. A merge combines the data of two pandas objects by finding matching values in one or more columns or row indexes. It then returns a new object that represents a combination of the data from both, based on relational-database-like join semantics applied to those values.

Merges are useful, as they allow us to model a single `DataFrame` for each type of data (one of the rules of having tidy data), but to be able to relate data in different `DataFrame` objects using values existing in both sets of data.

Merging data from multiple pandas objects

A practical example of merges would be that of looking up customer names from orders. To demonstrate this in pandas, we will use the following two `DataFrame` objects. One represents a list of customer details and the other represents the orders made by the customers and what day the order was made. They will be related to each other using the `CustomerID` columns in each:

```
In [21]:  # these are our customers
          customers = {'CustomerID': [10, 11],
                       'Name': ['Mike', 'Marcia'],
                       'Address': ['Address for Mike',
                                   'Address for Marcia']}
          customers = pd.DataFrame(customers)
          customers

Out[21]:                 Address  CustomerID    Name
          0     Address for Mike          10    Mike
          1   Address for Marcia          11  Marcia
```

```
In [22]:  # and these are the orders made by our customers
          # they are related to customers by CustomerID
          orders = {'CustomerID': [10, 11, 10],
                    'OrderDate': [date(2014, 12, 1),
                                  date(2014, 12, 1),
                                  date(2014, 12, 1)]}
          orders = pd.DataFrame(orders)
          orders

Out[22]:     CustomerID   OrderDate
          0          10   2014-12-01
          1          11   2014-12-01
          2          10   2014-12-01
```

Now, suppose we would like to ship the orders to the customers. We would need to merge the orders data with the customers detail data to determine the address for each order. This can be easily performed with the following statement:

```
In [23]:  # merge customers and orders so we can ship the items
          customers.merge(orders)

Out[23]:                 Address  CustomerID    Name    OrderDate
          0   Address for Mike          10    Mike   2014-12-01
          1   Address for Mike          10    Mike   2014-12-01
          2  Address for Marcia         11  Marcia   2014-12-01
```

pandas has done something magical for us here by being able to accomplish this with such a simple piece of code. It has realized that our customers and `orders` objects both have a column named `CustomerID` and, with this knowledge, it uses common values found in that column of both `DataFrame` objects to relate the data in both and form the merged data based on inner join semantics.

To be even more detailed on what occurs, what pandas specifically does is the following:

1. It determines the columns in both customers and orders with common labels. These columns are treated as the keys to perform the join.
2. It creates a new `DataFrame`, whose columns are the labels from the keys identified in the previous step, followed by all the non-key labels from both objects.
3. It matches values in the key columns of both `DataFrame` objects.
4. It then creates a row in the result for each set of matching labels.
5. It then copies the data from those matching rows from each source object into that respective row and columns of the result.
6. It assigns a new `Int64Index` to the result.

The join in a merge can use values from multiple columns. To demonstrate, the following creates two `DataFrame` objects and performs the merge using values in the `key1` and `key2` columns of both objects:

```
In [24]:  # data to be used in the remainder of this section's examples
          left_data = {'key1': ['a', 'b', 'c'],
                       'key2': ['x', 'y', 'z'],
                       'lval1': [ 0, 1, 2]}
          right_data = {'key1': ['a', 'b', 'c'],
                        'key2': ['x', 'a', 'z'],
                        'rval1': [ 6, 7, 8 ]}
          left = pd.DataFrame(left_data, index=[0, 1, 2])
          right = pd.DataFrame(right_data, index=[1, 2, 3])
          left

Out[24]:      key1 key2  lval1
          0    a    x       0
          1    b    y       1
          2    c    z       2
```

```
In [25]:  right

Out[25]:      key1 key2  rval1
          1    a    x       6
          2    b    a       7
          3    c    z       8
```

```
In [26]:  # demonstrate merge without specifying columns to merge
          # this will implicitly merge on all common columns
          left.merge(right)

Out[26]:      key1 key2  lval1  rval1
          0    a    x       0      6
          1    c    z       2      8
```

This merge identifies that the `key1` and `key2` columns are common in both the `DataFrame` objects. The matching tuples of values in both `DataFrame` objects for these columns are `(a, x)` and `(c, z)`, and therefore, this results in two rows of values.

To explicitly specify which column is used to relate the objects, you can use the `on` parameter. The following demonstrates this by performing a merge using only the values in the `key1` column of both the `DataFrame` objects:

```
In [27]:  # demonstrate merge using an explicit column
          # on needs the value to be in both DataFrame objects
          left.merge(right, on='key1')
```

```
Out[27]:    key1 key2_x  lval1 key2_y  rval1
          0    a      x      0      x      6
          1    b      y      1      a      7
          2    c      z      2      z      8
```

Comparing this result to the previous example, the result now has three rows, as there are matching a, b, and c values in that single column of both objects.

The on parameter can also be given a list of column names. The following reverts to using both the key1 and key2 columns, which results in an identical result to the previous example where those two columns were implicitly identified by pandas:

```
In [28]:  # merge explicitly using two columns
          left.merge(right, on=['key1', 'key2'])
```

```
Out[28]:    key1 key2  lval1  rval1
          0    a    x      0      6
          1    c    z      2      8
```

The columns specified with on need to exist in both the DataFrame objects. If you would like to merge based on columns with different names in each object, you can use the left_on and right_on parameters, passing the name or names of the columns to each respective parameter.

To perform a merge with the labels of the row indexes of the two DataFrame objects, you can use the left_index=True and right_index=True parameters (both need to be specified):

```
In [29]:  # join on the row indices of both matrices
          pd.merge(left, right, left_index=True, right_index=True)
```

```
Out[29]:    key1_x key2_x  lval1 key1_y key2_y  rval1
          1     b      y      1      a      x      6
          2     c      z      2      b      a      7
```

This has identified that the index labels in common are 1 and 2, so the resulting DataFrame has two rows with these values and labels in the index. pandas then creates a column in the result for every column in both the objects and then copies the values.

As both DataFrame objects had a column with identical name keys, the columns in the result have the _x and _y suffixes appended to them to identify the DataFrame object they originated from. _x is for left and _y is for right. You can specify these suffixes using the suffixes parameter and passing a two-item sequence.

Matrices and Linear Algebra

5

In this chapter, we will present the following recipes:

- Matrix operations and functions on two-dimensional arrays
- Solving linear systems using matrices
- Calculating the null space of a matrix
- Calculating the LU decompositions of a matrix
- Calculating the QR decomposition of a matrix
- Calculating the eigenvalue and eigenvector of a matrix
- Diagonalizing a matrix
- Calculating the Jordan form of a matrix
- Calculating the singular value decomposition of a matrix
- Creating a sparse matrix
- Computations on top of a sparse matrix

Introduction

In the previous chapters, we have become familiar with instantiating NumPy arrays, pandas data frames/series, and performing some basic data manipulation tasks using those. However, a typical data analysis task would involve a more detailed exploration, which in turn requires us to perform more scientific tasks.

This chapter discusses the solution of matrix-oriented problems in SciPy, which constitute the fundamentals of much of the work done in scientific computing and data analysis.

In order to understand the need for learning matrix analysis and linear algebra further, let us look into a few examples:

- **Image analysis**: Essentially, one can consider an image a matrix with m rows and n columns. In any type of image analysis, such as image classification or transformation, we can potentially work on the image by first converting it into a matrix format and then performing the analysis.
- **Optimization**: In some data analysis tasks, we could be optimizing certain parameters/variables—we might want to maximize or minimize those (for example, in a regression exercise, we would be minimizing the cost function). Linear algebra forms the basis of such optimization tasks.
- **Recommender systems**: One of the popular techniques used in building recommender systems is called singular value decomposition, and it is a combination of matrix analysis and linear algebra.
- **Text mining/market basket analysis**: There are multiple use cases in which the analyses results in a matrix where the majority of the matrix elements are zero, with only a few elements with a value of one. In the example of market basket analysis, if each transaction represents a row and each column represents a single product in the store, a majority of the rows would be zero, as the customer would be purchasing only a few out of the thousands of products present in the store. This is a classical example of a **sparse matrix** (a matrix with only a few ones). Learning to analyze on top of sparse matrices helps us code more efficiently.
- **Dimensionality reduction**: Extracting the eigenvalues and eigenvectors of a matrix is a key step in reducing the number of dimensions (number of columns in a dataset) without losing a lot of information from the dataset.

Matrix operations and functions on two-dimensional arrays

Basic matrix operations form the backbone of quite a few statistical analyses—for example, neural networks. In this section, we will be covering some of the most used operations and functions on 2D arrays:

- Addition
- Multiplication by scalar
- Matrix arithmetic

- Matrix-matrix multiplication
- Matrix inversion
- Matrix transposition

In the following sections, we will look into the methods of implementing each of them in Python using SciPy/NumPy.

How to do it...

Let's look the the different methods.

Matrix addition

In order to understand how matrix addition is done, we will first initialize two arrays:

```
# Initializing an array
x = np.array([[1, 1], [2, 2]])
y = np.array([[10, 10], [20, 20]])
```

Similar to what we saw in a previous chapter, we initialize a 2 x 2 array by using the `np.array` function.

There are two methods by which we can add two arrays.

Method 1

A simple addition of the two arrays x and y can be performed as follows:

```
x+y
```

Note that x evaluates to:

```
[[1 1]
 [2 2]]
```

y evaluates to:

```
[[10 10]
 [20 20]]
```

The result of x+y would be equal to:

```
[[1+10 1+10]
 [2+20 2+20]]
```

Finally, this gets evaluated to:

```
[[11 11]
 [22 22]]
```

Method 2

The same preceding operation can also be performed by using the add function in the numpy package as follows:

```
np.add(x,y)
```

Multiplication by a scalar

Matrix multiplication by a scalar can be performed by multiplying the vector with a number. We will perform the same using the following two steps:

1. Initialize a two-dimensional array.
2. Multiply the two-dimensional array with a scalar.

We perform the steps, as follows:

1. To initialize a two-dimensional array:

   ```
   x = np.array([[1, 1], [2, 2]])
   ```

2. To multiply the two-dimensional array with the k scalar:

   ```
   k*x
   ```

For example, if the scalar value k = 2, then the value of k*x translates to:

```
2*x
array([[2, 2],
       [4, 4]])
```

Matrix arithmetic

Standard arithmetic operators can be performed on top of NumPy arrays too. The operations used most often are:

- Addition
- Subtraction

- Multiplication
- Division
- Exponentials

The other major arithmetic operations are similar to the addition operation we performed on two matrices in the *Matrix addition* section earlier:

```
# subtraction
x-y
array([[ -9,  -9],
       [-18, -18]])
# multiplication
x*y
array([[10, 10],
       [40, 40]])
```

While performing multiplication here, there is an element to element multiplication between the two matrices and not a matrix multiplication (more on matrix multiplication in the next section):

```
# division
x/y
array([[ 0.1,  0.1],
       [ 0.1,  0.1]])
# exponential
x**y
array([[       1,        1],
       [1048576, 1048576]], dtype=int32)
```

Matrix-matrix multiplication

Matrix to matrix multiplication works in the following way:

1. We have a set of two matrices with the following shape:

$$A = \begin{pmatrix} A_11 & A_12 & \cdots & A_1m \\ A_21 & A_22 & \cdots & A_2m \\ \vdots & \vdots & \vdots & \vdots \\ A_n1 & A_n2 & \cdots & A_nm \end{pmatrix}, B = \begin{pmatrix} B_11 & B_12 & \cdots & B_1m \\ B_21 & B_22 & \cdots & B_2m \\ \vdots & \vdots & \vdots & \vdots \\ B_n1 & B_n2 & \cdots & B_nm \end{pmatrix}$$

Matrix *A* has *n* rows and *m* columns and matrix *B* has *m* rows and *p* columns.

2. The matrix multiplication of *A* and *B* is calculated as follows:

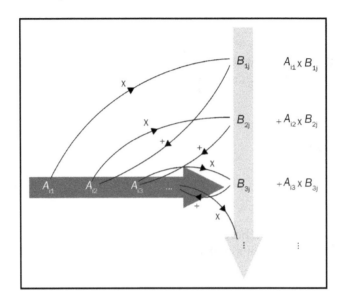

The matrix operation is performed by using the built-in `dot` function available in NumPy as follows:

1. Initialize the arrays:

```
x=np.array([[1, 1], [2, 2]])
y=np.array([[10, 10], [20, 20]])
```

2. Perform the matrix multiplication using the `dot` function in the numpy package:

```
np.dot(x,y)
array([[30, 30],
       [60, 60]])
```

3. The np.dot function does the multiplication in the following way:

```
array([[1*10 + 1*20, 1*10 + 1*20],
       [2*10 + 2*20, 2*10 + 2*20]])
```

Whenever matrix multiplication happens, the number of columns in the first matrix should be equal to the number of rows in the second matrix.

Matrix transposition

Matrix transposition is performed by using the `transpose` function available in `numpy` package.

The process to generate the transpose of a matrix is as follows:

1. Initialize a matrix:

   ```
   A = np.array([[1,2],[3,4]])
   ```

2. Calculate the transpose of the matrix:

   ```
   A.transpose()
   array([[1, 3],
          [2, 4]])
   ```

 The transpose of a matrix with *m* rows and *n* columns would be a matrix with *n* rows and *m* columns

Matrix inversion

While we performed most of the basic arithmetic operations on top of matrices earlier, we have not performed any specialist functions within scientific computing/analysis—for example, matrix inversion, transposition, ranking of a matrix, and so on.

The other functions available within the `scipy` package shine through (over and above the previously discussed functions) in such a scenario where more data manipulation is required apart from the standard ones.

Matrix inversion can be performed by using the function available in `scipy.linalg`. The process to perform matrix inversion and its implementation in Python is as follows:

1. Import relevant packages and classes/functions within a package:

   ```
   from scipy import linalg
   ```

2. Initialize a matrix:

   ```
   A = np.array([[1,2],[3,4]])
   ```

3. Pass the initialized matrix through the `inverse` function in package:

   ```
   linalg.inv(A)
   array([[-2. ,  1. ],
          [ 1.5, -0.5]])
   ```

Solving linear systems using matrices

A typical linear equation is expressed in the form of *ax+by=c*, where *x* and *y* are unknowns and *a*, *b*, and *c* are known values. Essentially, we are trying to find a relationship between *a*, *b*, and *c*.

There are multiple ways in which one is able to solve linear equations. In this recipe, we will be looking into the built-in function available in SciPy that helps us in solving a given linear system.

How it works...

The intuition behind solving for *x* and *y* is as follows:

$$([[a1\,, b1\,],\quad x\quad ([x],\quad =\quad ([c1],$$
$$[a2\,, b2\,]])\qquad [y])\qquad [c2])$$

For convenience, let's call the left-hand side matrix *(a1, b1, a2, b2)* the input matrix and the right-hand side matrix *(c1, c2)* the right-hand side vector.

The preceding matrix multiplication translates to:

$$a1 * x + b1 * y = c1$$
$$a2 * x + b2 * y = c2$$

Given that there are two equations and two unknowns (*x* and y), we are in a position to solve for *x* and *y*. We should notice that, given that the matrix has two rows and columns with two unknowns, we can solve the unknowns.

More generally (unless in some special cases, which we will go through later), if we have a matrix with *n* columns, we will have *n* variables to solve for.

Solving linear systems of equations is straightforward using the SciPy command `linalg.solve`. This command expects an input matrix and an output vector. The solution vector (x and y) is then computed. For example, suppose we want to solve the following simultaneous equations:

$$x + 3y + 5z = 10$$
$$2x + 5y + z = 8$$
$$2x + 3y + 8z = 3$$

We can translate the preceding set of equations into a matrix form, as follows:

$$
\begin{array}{cccc}
[[1,3,5], & [[x], & & [[10], \\
[2,5,1], & X \quad [y], & = & [8], \\
[2,3,8]] & [z]] & & [3]]
\end{array}
$$

Solving for x, y, and z is done as follows:

$$
\begin{bmatrix} x \\ y \\ z \end{bmatrix} = \begin{bmatrix} 1 & 3 & 5 \\ 2 & 5 & 1 \\ 2 & 3 & 8 \end{bmatrix}^{-1} \begin{bmatrix} 10 \\ 8 \\ 3 \end{bmatrix} = \frac{1}{25} \begin{bmatrix} -232 \\ 129 \\ 19 \end{bmatrix} = \begin{bmatrix} -9.28 \\ 5.16 \\ 0.76 \end{bmatrix}
$$

-1 on top stands for the inverse of a matrix.

How to do it...

The whole preceding process is implemented in code as follows:

1. Import relevant classes within a package:

```
from scipy import linalg
```

2. Initialize the right-hand side and left-hand side matrices (input and output matrices):

```
A=[[1,3,5],[2,5,1],[2,3,8]]
b=[[10],[8],[3]]
```

Now, we will look into the function used to solve the equation $A*x = b$.

Method 1 (using the functions we have learnt so far)

Multiply the inverse of A with b.

```
np.dot(linalg.inv(A),b)
```

Method 2 (using the solve function)

Using the `solve` function within the `scipy` package helps in solving the equation straight away:

```
scipy.linalg.solve(x, b)
```

The output of either of the preceding methods is the solution, as follows:

```
array([[-9.28],       [ 5.16],       [ 0.76]])
```

Calculating the determinant of a matrix

As seen earlier, in order to solve a linear system, we need to calculate the inverse of a matrix first, which in turn requires us to calculate the determinant of the matrix.

The way in which the determinant is calculated is as follows:

$$A = \begin{bmatrix} 1 & 3 & 5 \\ 2 & 5 & 1 \\ 2 & 3 & 8 \end{bmatrix}$$

$$|A| = 1 \begin{vmatrix} 5 & 1 \\ 3 & 8 \end{vmatrix} - 3 \begin{vmatrix} 2 & 1 \\ 2 & 8 \end{vmatrix} + 5 \begin{vmatrix} 2 & 5 \\ 2 & 3 \end{vmatrix}$$

$$= 1(5 \cdot 8 - 3 \cdot 1) - 3(2 \cdot 8 - 2 \cdot 1) + 5(2 \cdot 3 - 2 \cdot 5) = -25$$

 $|A|$ stands for the determinant of a matrix.

The determinant of a matrix in Python can be calculated by using the `det` function in `scipy.linalg`.

The same preceding calculation can be implemented in code, as follows:

```
linalg.det(A)
```

Calculating the null space of a matrix

The null space of an *m x n* matrix *A*, denoted as *null A*, is the set of all solutions for the homogeneous equation *Ax = 0*.

Calculating the null space of a matrix helps us in identifying all the potential values of *x* that help solve the equation *Ax = 0*.

In order to calculate the null space of a given matrix, we would be using the built-in `nullspace` function available within the `sympy` package.

In order to understand how the null space of a given matrix can be calculated, let us consider the following example:

1. Initialize the matrix:

   ```
   M = Matrix([[1, 2, 3, 0, 0], [4, 10, 0, 0, 1]])
   ```

2. Calculate the null space of the matrix by using the `nullspace` function:

   ```
   M.nullspace()
   ```

3. The output of the preceding code is:

   ```
   [Matrix([[-15],[  6],[  1],[  0],[  0]]),
   Matrix([[0],[0],[0],[1],[0]]), Matrix([[   1],[-1/2],[   0],[
   0],[   1]])]
   ```

   In order to cross-check the preceding output, let us perform a matrix multiplication of each of the preceding outputs with the original matrix; the expected output will be a value of 0.

4. Let's initialize one of the preceding outputs:

   ```
   x=[[-15],[  6],[  1],[  0],[  0]]
   ```

5. The matrix multiplication of M and x will be performed as follows:

   ```
   np.dot(np.array(M),x)
   ```

6. The output of the preceding code is:

   ```
   array([[0],        [0]], dtype=object)
   ```

The preceding code validates that the output of the preceding matrix multiplication is a zero matrix and hence the nullspace solutions obtained are valid.

Calculating the range of a matrix

The range of a set of elements is the difference between the maximum and minimum values present in the list of elements.

In a typical analysis, calculating the range of elements helps in scaling the variables properly. For example, min-max scaling uses the minimum and maximum value of a set of elements in order to scale them.

For a given matrix, we would be able to calculate the range of the matrix in multiple ways, as follows:

- Range of the whole matrix
- Range of the elements in the rows of a matrix
- Range of the elements in the columns of a matrix

Range of the whole matrix

In order to calculate the range of a matrix, we use the `ptp` function within the `numpy` package in Python.

ptp stands for **peak to peak**.

Let us look into the workings of the function with an example.

1. Initialize `matrixx=np.array([[1,100],[2,200]])`.
2. Calculate the range of the whole matrix:

```
np.ptp(x)
100
```

3. The output of the preceding function is `199`, as the maximum element's value is `200`, while the minimum value is `1`.

Range of elements in the rows of a matrix

Rows and columns of a matrix can be specified by using the `axis` parameter.

The following function calculates the range of rows of a matrix:

```
np.ptp(x,axis=1)
array([ 99, 198])
```

It is to be noted that the range in the first row of the matrix is `99`, as the maximum value is `100` and the minimum value is `1`. The range of second row calculated is similar.

Range of elements in the columns of a matrix

Similar to the range calculation of rows, the range calculation of columns can be calculated by specifying `axis=0` using the same function:

```
np.ptp(x,axis=0)
array([  1, 100])
```

Calculating the rank and nullity of a matrix

The rank of a matrix is defined as the maximum number of linearly independent column vectors in the matrix or the maximum number of linearly independent row vectors in the matrix.

In order to understand how the rank of a matrix is calculated, we will look into the following example:

$$\begin{bmatrix} 3 & 2 & -1 \\ 2 & -3 & -5 \\ -1 & -4 & -3 \end{bmatrix}$$

Let's reorder the rows in such a way that the third row moves to the top and the first rows are shifted by one row. The resulting matrix is as follows:

$$\begin{bmatrix} 1 & 4 & 3 \\ 3 & 2 & -1 \\ 2 & -3 & -5 \end{bmatrix}$$

We'll transform the preceding matrix in such a way that the third row would be transformed into a row of zeros. We can achieve that by first multiplying the second row with a value of 3 and then multiplying the third row with a value of 2.

Similarly, we will multiply the second row by a value of 3 and subtract from the first row. The result of the preceding transformation would be as follows:

$$\begin{bmatrix} 1 & 4 & 3 \\ 0 & -10 & -10 \\ 0 & -11 & -11 \end{bmatrix}$$

From the preceding output, we will divide the second row by a value of -10. The resulting matrix is as follows:

$$\begin{bmatrix} 1 & 4 & 3 \\ 0 & 1 & 1 \\ 0 & -11 & -11 \end{bmatrix}$$

Finally, we add the third row with 11 times that of second row. This transformation results in a matrix of the form as follows:

$$\begin{bmatrix} 1 & 4 & 3 \\ 0 & 1 & 1 \\ 0 & 0 & 0 \end{bmatrix}$$

Given that two rows out of the three rows have at least one non-zero element in the matrix, the rank of the matrix is 2.

How to do it...

In the following section, we will look into how to implement the same in Python.

The function to extract the matrix rank of a given matrix is available in numpy.linalg as matrix_rank.

1. Import the relevant packages:

```
import numpy as np
```

2. Initialize a matrix:

```
a=np.array([[3,2,-1],[2,-3,-5],[-1,-4,-3]])
```

3. Use the relevant function:

```
np.linalg.matrix_rank(a)
2
```

From the preceding code, we can see that the rank of the initialized matrix is 2.

The nullity of a matrix is the difference between the number of columns and the rank of a matrix.

In this particular example, the nullity of matrix is as follows:

$$3 - 2 = 1$$

In the following code, we'll look into how to calculate the nullity of a matrix:

1. Calculate the number of columns of a matrix:

```
a.shape[1]
```

2. Calculate the nullity of a matrix:

```
np.linalg.matrix_rank(a) - a.shape[1]
1
```

Calculating the LU decompositions of a matrix

LU decomposition (where **LU** stands for **lower upper**, also called **LU factorization**) factors a matrix as the product of a lower triangular matrix and an upper triangular matrix, such that the product of these two matrices provides the original matrix.

This method of factorizing a matrix as a product of two triangular matrices having various applications such as finding the solution of a system of equations, which itself is an integral part of many applications, finding the current in a circuit and solution of discrete dynamical system problems, finding the inverse of a matrix, and finding the determinant of the matrix.

Basically, the LU decomposition method comes in handy whenever it is possible to model the problem to be solved in the matrix form. Conversion to the matrix form and solving with triangular matrices makes it easy to do calculations in the process of finding the solution.

A square matrix A can be decomposed into two square matrices L and U such that $A = L\,U$ where U is an upper triangular matrix formed as a result of applying Gauss elimination method on A. L is a lower triangular matrix with diagonal elements being equal to 1:

$$For\,A = \begin{bmatrix} a_{11} & a_{12} & a_{13} \\ a_{21} & a_{22} & a_{23} \\ a_{31} & a_{32} & a_{33} \end{bmatrix},\, we\ have\ L = \begin{bmatrix} 1 & 0 & 0 \\ l_{21} & 1 & 0 \\ l_{31} & l_{32} & 1 \end{bmatrix}\ and\ U = \begin{bmatrix} u_{11} & u_{12} & u_{13} \\ 0 & u_{22} & u_{23} \\ 0 & 0 & u_{33} \end{bmatrix};\ such\ that\ A = LU$$

How to do it...

The function that is useful in factorizing a matrix into a lower and upper triangular matrix is available in `scipy.linalg` as the `lu` function.

In order to understand how LU factorization works, let us go through the following code snippets:

1. Define a matrix:

   ```
   A = np.array([[2., 1., 1.], [1., 3., 2.], [1., 0., 0.]])
   ```

2. Import relevant classes and packages:

   ```
   import scipy.linalg as linalg
   ```

3. Calculate the LU matrices using the `lu` function in `scipy.linalg`:

   ```
   P, L, U = scipy.linalg.lu(A)
   ```

4. Check the output of L:

   ```
   print(L)
   [[ 1.    0.    0. ] [ 0.5   1.    0. ] [ 0.5  -0.2   1. ]]
   ```

5. Check the output of U`print(U)`:

   ```
   [[ 2.    1.    1. ] [ 0.    2.5   1.5] [ 0.    0.   -0.2]]
   ```

Note that, in the preceding outputs, L is a lower triangular matrix (all the elements in the upper half of the matrix are 0) and U is an upper triangular matrix.

Once we obtain the L and U matrices, let us cross-check if the matrix multiplication of the two results in the input matrix A.

The matrix multiplication of *L, U* is as follows:

```
np.dot(L,U)
[[  2.00000000e+00    1.00000000e+00    1.00000000e+00]
 [  1.00000000e+00    3.00000000e+00    2.00000000e+00]
 [  1.00000000e+00   -2.77555756e-17    0.00000000e+00]]
```

In the preceding output, we notice that it is the same as the input matrix *A*, which we created earlier.

Calculating the QR decomposition of a matrix

Similar to the LU decomposition, QR decomposition is the decomposition of an original matrix into its constituent parts.

In this particular case, the matrix *A = QR*, where *Q* is an orthogonal matrix and *R* is an upper triangular matrix.

Before getting into further details, let us look into the properties of an orthogonal matrix:

- It is a square matrix
- Multiplying *Q* with its transpose results in an identity matrix

How to do it...

QR decomposition can be done by using the `qr` function within `scipy.linalg`.

In the following code, let us look into how QR decomposition works:

1. Load the relevant packages:

    ```
    import scipy.linalg as linalg
    import numpy as np
    ```

2. Initialize a matrix:

    ```
    A = np.array([ [2., 1., 1.], [1., 3., 2.], [1., 0., 0]])
    ```

3. Extract the Q and R matrices of the matrix A:

```
Q, R = linalg.qr(A)
```

4. Check the output values:

```
print(Q)
[[-0.81649658  0.27602622 -0.50709255]
 [-0.40824829 -0.89708523  0.16903085]
 [-0.40824829  0.34503278  0.84515425]]
print(R)
[[-2.44948974 -2.04124145 -1.63299316]
 [ 0.         -2.41522946 -1.51814423]
 [ 0.          0.         -0.16903085]]
```

Once we have the results, let us look into the preceding properties mentioned.

1. Multiplying Q with its transpose results in an identity matrix:

```
print(np.dot(Q,Q.T))
[[  1.00000000e+00  -1.05102508e-17  -3.90671821e-17]
 [ -1.05102508e-17   1.00000000e+00   2.51394744e-17]
 [ -3.90671821e-17   2.51394744e-17   1.00000000e+00]]
```

We can see that, apart from the diagonal elements, the other elements are very small and are close to zero, thus resulting in an identity matrix .

2. Matrix multiplication of Q and R would result in A:

```
print(np.dot(Q,R))
[[  2.00000000e+00   1.00000000e+00   1.00000000e+00]
 [  1.00000000e+00   3.00000000e+00   2.00000000e+00]
 [  1.00000000e+00   1.09812324e-16  -2.08416416e-17]]
```

We can see that the matrix multiplication of Q and R is indeed A.

Calculating the eigenvalue and eigenvector of a matrix

One of the major advantages of eigenvalue calculation is its ability to reduce the dimensions of a dataset, which in turn reduces the computations required to solve a given set of variables.

The eigenvector of a given vector is the vector that satisfies the following condition:

$$A \bullet v = \lambda v$$

In the preceding equation, A is the matrix of our interest, v is the eigenvector, and λ is the eigenvalue of the given matrix.

How to do it...

In SciPy, we calculate the eigenvector and eigenvalue of a given matrix by using the eig function in `scipy.linalg`.

Using the following code, let us look at calculating the eigenvector and the corresponding eigenvalue of a given matrix:

1. Initialize a matrix:

   ```
   a = np.array([[1, 2], [3, 4]])
   ```

2. Calculate the eigenvalue and eigenvector of the matrix:

   ```
   la, v = linalg.eig(a)
   ```

3. The output of the preceding code is:

   ```
   print(la)
   [-0.37228132+0.j   5.37228132+0.j]
   print(v)
   [[-0.82456484 -0.41597356] [ 0.56576746 -0.90937671]]
   ```

Note that la is the eigenvalue and the v matrix is the eigenvector.

Let us cross-check the preceding output based on the intuition we laid out at the start of this section.

The eigenvector for a given matrix A satisfies the following criterion:

$$A \bullet v = \lambda v$$

The following applies:

- *A* is the matrix
- *v* is the eigenvector
- λ is the eigenvalue

Let us cross-check the output based on the intuition we laid out earlier.

1. Calculate the value of the matrix multiplication of *A•v*:

```
np.dot(a,v)
array([[ 0.30697009, -2.23472698],          [-0.21062466,
-4.88542751]])
```

2. Calculate the value of the matrix multiplication of $\lambda\, v$:

```
la*v
array([[ 0.30697009-0.j, -2.23472698+0.j],
       [-0.21062466+0.j, -4.88542751+0.j]])
```

Note that the output satisfies the initially laid out equation: *A•v* = $\lambda\, v$.

Diagonalizing a matrix

In linear algebra, a square matrix *A* is diagonalizable if it is similar to a diagonal matrix, that is, if there exists an invertible matrix *P* such that $P^{-1}AP$ is a diagonal matrix.

Diagonalizable matrices and maps are of interest because diagonal matrices are especially easy to handle. If their eigenvalues and eigenvectors are known, one can raise a diagonal matrix to a power by simply raising the diagonal entries to that same power, and the determinant of a diagonal matrix is simply the product of all diagonal entries.

How to do it...

Before getting into the details of checking whether a matrix is diagonalizable or not, let us see how to convert any 1D array into a diagonal matrix.

1. Import the relevant packages:

```
import numpy as np
```

2. Initialize a 1D matrix:

```
a= np.array([1,2,3])
```

3. Diagonalize the matrix:

```
np.diag(a)
```

4. The output of the preceding code is:

```
array([[1, 0, 0],        [0, 2, 0],        [0, 0, 3]])
```

You should note that all the elements within the one-dimensional array form the diagonal of the output matrix generated by the diag function within the numpy package. Moreover, the other elements of the output array (apart from the diagonal elements) are 0. In the following code, we will look into checking whether a matrix is diagonalizable or not.

As discussed earlier, a matrix A is diagonalizable if there exists another matrix P that satisfies the following condition:

$$P^{-1}AP \text{ is a diagonal matrix}$$

In this example, we will consider the eigenvector of the given matrix as our P matrix.

1. We can calculate the P matrix as follows:

```
la, v = linalg.eig(a)
print(la)
print(v)
[ 3.+0.j -1.+0.j -3.+0.j]
[[ 0.70710678 -0.70710678  0.        ]
 [ 0.70710678  0.70710678  0.        ]
 [ 0.          0.          1.        ]]
```

As seen earlier in the eigenvector calculation, the matrix v would be the eigenvector (which is the P matrix for us).

2. Once we have the P matrix, $P^{-1}AP$ is calculated as follows:

```
np.dot(np.dot(np.linalg.inv(v),a),v)
```

The output of preceding code is:

```
array([[  3.00000000e+00,   5.05623488e-16,   0.00000000e+00],
 [  6.57231935e-16,  -1.00000000e+00,   0.00000000e+00],
 [  0.00000000e+00,   0.00000000e+00,  -3.00000000e+00]])
```

3. Diagonalizing the eigenvalues of our matrix is done as follows:

```
np.diag(la)
```

The output of preceding code will be:

```
array([[ 3.+0.j,  0.+0.j,  0.+0.j],
       [ 0.+0.j, -1.+0.j,  0.+0.j],
       [ 0.+0.j,  0.+0.j, -3.+0.j]])
```

4. Compare the difference between the outputs in step 2 and step 3:

```
np.dot(np.dot(np.linalg.inv(v),a),v)-np.diag(la)
array([[ -4.44089210e-16+0.j,   5.05623488e-16+0.j,
0.00000000e+00+0.j],
       [  6.57231935e-16+0.j,  -3.33066907e-16+0.j,   0.00000000e+00+0.j],
       [  0.00000000e+00+0.j,   0.00000000e+00+0.j,
0.00000000e+00+0.j]])
```

In the preceding calculation, note that the difference between the two outputs is very small and, hence, satisfies our originally laid out objective; $P^{-1}AP$ is a diagonal matrix.

Note that the diagonal elements within the diagonal matrix are the eigenvalues of the original matrix.

Calculating the Jordan form of a matrix

A Jordan block with value λ is a square, upper triangular matrix whose entries are all λ on the diagonal, 1 on the entries immediately above the diagonal, and 0 elsewhere.

Typical Jordan blocks of size *1, 2,* and *3* look like the following:

$$[\lambda], \quad \begin{bmatrix} \lambda & 1 \\ 0 & \lambda \end{bmatrix}, \quad \begin{bmatrix} \lambda & 1 & 0 \\ 0 & \lambda & 1 \\ 0 & 0 & \lambda \end{bmatrix}$$

A Jordan form matrix is a block diagonal matrix whose blocks are all Jordan blocks. For example, the following matrices are all Jordan form matrices:

$$
\begin{bmatrix}
1 & 1 & 0 & 0 & 0 & 0 \\
0 & 1 & 0 & 0 & 0 & 0 \\
0 & 0 & 3 & 1 & 0 & 0 \\
0 & 0 & 0 & 3 & 0 & 0 \\
0 & 0 & 0 & 0 & -1 & 0 \\
0 & 0 & 0 & 0 & 0 & -1
\end{bmatrix}
\qquad
\begin{bmatrix}
2 & 1 & 0 & 0 \\
0 & 2 & 1 & 0 \\
0 & 0 & 2 & 0 \\
0 & 0 & 0 & 2
\end{bmatrix}
\qquad
\begin{bmatrix}
2 & 1 & 0 & 0 \\
0 & 2 & 0 & 0 \\
0 & 0 & 2 & 1 \\
0 & 0 & 0 & 2
\end{bmatrix}
$$

In the first matrix among the preceding set of matrices, we should notice that the block on the top left *[[1,1],[0,1]]* is a Jordan block of size 2, while the matrix on the bottom right is a Jordan block of size 1.

How to do it...

In order to calculate the Jordan form of a matrix, we will using the `jordan_form` function available within the `sympy` package.

In the following code, we will look into obtaining the Jordan form of a given matrix in Python.

1. Import the relevant packages:

```
import numpy as np
from sympy import Matrix
```

2. Initialize an array of numbers:

```
a = np.array([[5, 4, 2, 1], [0, 1, -1, -1], [-1, -1, 3, 0], [1, 1,
-1, 2]])
```

3. Convert the array into a matrix format to be used by the `jordan_form` function:

```
m = Matrix(a)
```

4. Apply the function on the preceding matrix:

```
P, J = m.jordan_form()
```

5. Extract the output of the `jordan_form`, which is `J`:

```
J
Matrix([[1.0,    0,    0,    0],[ 0, 2.0,    0,    0],[ 0,    0, 4.0,
1],[ 0,    0,    0, 4.0]])
```

Calculating the singular value decomposition of a matrix

Singular value decomposition (**SVD**) is one of the more useful techniques in typical data science techniques.

- One of the most important applications of SVD is in recommendation systems, where the matrix of user-item purchase behavior is broken into multiple matrices that are simpler to implement.
- Similarly, SVD is used in image compression algorithms, where we try to capture the information within algorithms by using as few pixels as possible.

 The SVD of a matrix A is the decomposition or factorization of A into the product of three matrices: $A=Ux\Sigma xVt$.

The size of the individual matrices is as follows, if you know that matrix A is of size M x N:

- Matrix U is of size M x M
- Matrix V is of size N x N
- Matrix Σ is of size M x N

How to do it...

The SVD of a matrix can be calculated by using the `svd` function within `scipy.linalg`.

Let us understand how to code up using the `svd` function using the following example:

1. Initialize an array:

```
a = np.array([[1, 2], [3, 4]])
```

2. Import the relevant packages and functions within the package:

```
from scipy import linalg
```

3. Use the `svd` function to calculate the SVD matrices (*U*, *Σ* and *V*):

```
linalg.svd(a)
```

4. The output of the preceding function is as follows:

```
(array([[-0.40455358, -0.9145143 ],
        [-0.9145143 ,  0.40455358]]),
 array([ 5.4649857 ,  0.36596619]),
 array([[-0.57604844, -0.81741556],
        [ 0.81741556, -0.57604844]]))
```

In order to further understand if the output is as per expectation, let us check if the multiplication of the individual matrices in the output results in a matrix that is very similar to our original matrix.

In this scenario, our *U* matrix is as follows:

```
([[-0.40455358,-0.9145143],[-0.9145143,0.40455358]])
```

The *Σ* matrix is as follows:

```
([ 5.4649857 ,  0.36596619])
```

However, note that the preceding second output of the `svd` function is only the diagonal elements within the matrix (where every non-diagonal element is 0). Hence, the *Σ* matrix translates to the following:

```
np.array([[ 5.4649857 ,  0],
          [0,0.36596619]])
```

Note that the diagonal values are populated by the elements of the `svd` output we saw earlier.

The *V* matrix is as follows:

```
([[-0.57604844, -0.81741556],  [ 0.81741556, -0.57604844]])
```

In order to cross-check and verify the `svd` function's output, we will multiply the *U*, *Σ*, and *V* matrices that we obtained earlier.

The output of the matrix multiplication should be as close to the original matrix as possible.

1. Initialize the matrices:

```
U = np.array([[-0.40455358, -0.9145143 ],
        [-0.9145143 ,  0.40455358]])
sigma = np.array([[ 5.4649857 ,   0],
             [0,0.36596619]])
V=np.array([[-0.57604844, -0.81741556],
        [ 0.81741556, -0.57604844]])
```

2. Perform the matrix multiplication of *U*, sigma, and *V* matrices:

```
np.dot(np.dot(U,sigma),V)
```

3. The output of the preceding calculation is as follows:

```
array([[ 0.99999999,  1.99999998],
        [ 3.00000003,  4.00000001]])
```

Note that, the output matrix is almost the same as the original matrix.

Creating a sparse matrix

In order to understand sparse matrices, we will consider the following real-world scenario: recommending the next item that a supermarket customer is likely to buy, given a set of historical transactions.

In a typical supermarket, there can be millions of customers and thousands of items. Any given user would have bought only a few items among the thousands of items present in the supermarket.

We can represent all the transactions of a supermarket in such a way that all the customers are represented in rows and all the items are represented in columns. The cell values are 1 if the customer bought the item, and 0 otherwise.

In the preceding scenario, we will have a very high majority of zeros and very few ones. This scenario, where the number of ones is extremely low, is called sparsity (sparse number of ones). Hence, the matrix is called a sparse matrix.

The need for a better representation of the sparse matrix

In order to understand the need for representing a matrix as sparse matrix, let us look at the following scenario.

There are a million users of a supermarket and 10,000 items available in it. If we were to represent the transactions, it would be a 1,000,000 x 10,000 dimensional matrix, where the number of rows is the number of customers and the number of columns is the number of items in the matrix.

The number of 1s in the preceding matrix will be extremely sparse, so it would be a better idea to store only the index locations where the indices are 1 and everything else is 0.

For example, in a hypothetical case, where every customer has bought only five items in the supermarket, there would be five million ones (one million customers and five items bought per customer).

If we were representing the matrix in the traditional format, there would be 1,000,000 x 10,000 = 10 billion numbers that need to be stored. However, by storing the indices of the ones, we are only storing five million numbers in place of 10 billion numbers. Thus, sparse matrices help in speeding up our calculations, given that they work with a lot less data than the original matrix.

In the following sections, we will look into building a sparse matrix in Python.

How to do it...

Converting a matrix into its sparse form

Sparse matrices are built using the functions available in `scipy.sparse`.

In the following example, we will be using the function named `csr_matrix` to convert a given matrix into its sparse form.

1. Import the relevant packages:

```
import numpy as np
from scipy import sparse
```

2. Initialize a matrix:

```
A = np.array([[1,2,0],[0,0,3],[1,0,4]])
```

3. Convert the initialized matrix into a sparse matrix:

```
sA = sparse.csr_matrix(A)
```

4. Check the output of the preceding sparse matrix initialization:

```
sA
<3x3 sparse matrix of type '<class 'numpy.int32'>'        with 5
stored elements in Compressed Sparse Row format>
```

Note that the preceding output mentions that there are five stored elements, which is the same as having five non-zero elements within the initialized matrix.

5. Print the output of sA:

```
print(sA)
    (0, 0)        1
    (0, 1)        2
    (1, 2)        3
    (2, 0)        1
    (2, 2)        4
```

The left-hand side of the preceding output indicates the row and column indices of the original matrix, where there is a non-zero element. The right-hand side of the preceding matrix provides the values of the elements located in the given indices.

Thus, storing elements in this way would require much less space when the matrix is huge and a majority of the elements are zero.

Different ways of creating a sparse matrix

In the previous section, we looked into a function that creates sparse matrices, called, csr_matrix. **csr_matrix** stands for **compressed sparse row matrix**. Similar to a csr matrix, there are multiple other formats of sparse matrices. The following table gives a list of all the multiple formats of sparse matrices:

Bsr_matrix	**Block sparse row matrix**
Coo_matrix	Sparse matrix in coordinate format
Csc_matrix	Compressed sparse column matrix
Csr_matrix	Compressed sparse row matrix
Dia_matrix	Sparse matrix with diagonal storage
Dok_matrix	Dictionary of keys based sparse matrix

`Lil_matrix`	Row-based linked list sparse matrix

Computations on top of a sparse matrix

In order to understand how to perform computations on top of a sparse matrix and the resulting benefits thereof, we will be looking at an example and comparing the difference between having a sparse matrix and not having a sparse matrix.

Solving a system of equations

As discussed in the *Solving linear systems using matrices* recipe, a system of equations is solved using the `solve` function in `scipy.linalg`.

In order to compare the difference between sparse matrix computation and non-sparse matrix computation, we will perform the following tasks:

- Import relevant packages
- Initialize a 10,000 x 10,000 matrix named *A*
- Impute very few values with some random numbers
- Set the diagonal, so that the rank of matrix is not reduced by a lot
- Initialize a set of values for the output *b* so that the equation $A*x = b$ is set up
- Solve for the values of *x* once with the sparse matrix and another time without the sparse matrix format
- Compare the differences in the speed of computation by using the sparse matrix format

How to do it...

Lets get into the details.

1. Import the relevant packages:

```
from scipy.sparse import lil_matrix
from scipy.sparse.linalg
import spsolve
from numpy.linalg import solve, norm
from numpy.random import rand
```

2. Initialize a matrix:

```
A = lil_matrix((10000, 10000))
```

In the preceding snippet of code, we have initialized a linked list matrix that is 10,000 x 10,000 in size.

3. Impute very few values with random numbers:

```
A[0, :100] = rand(100)
A[1, 100:200] = A[0, :100]
```

In the preceding snippet of code, we have imputed the first 100 columns of the first row of the matrix, and the next 100 columns of the second row of the matrix with random numbers.

4. Set the diagonal of the original matrix:

```
A.setdiag(rand(10000))
```

In the preceding set of code, we used the `setdiag` function to initialize the diagonal values with a random set of numbers. With all the preceding steps complete, we have successfully initialized the input matrix.

5. Initialize the output:

```
matrixb = rand(10000)
```

In the preceding line of code, we have initialized b to a random set of 10,000 number outputs.

6. Convert the input matrix into `csr` format, as the function used to *solve* the equation works only on `csr` or `csc` matrices:

```
A = A.tocsr()
```

In the preceding snippet of code, we have converted the linked list matrix format into `csr` format.

7. Solve for the equation $Ax = b$:

```
x = spsolve(A, b)
```

Note that, in order to solve for the equation, we used the `spsolve` function instead of the `solve` function that we learnt about earlier.

In order to understand the efficiency gained by using a sparse matrix over a dense one, we will compare the time taken to solve using a sparse matrix and the time taken to compute using a dense matrix.

We will be using the functions available in the `time` package to compute the time taken to execute a snippet of code.

1. Import the relevant packages:

   ```
   import time
   ```

2. Calculate the time taken to compute using the sparse matrix format:

   ```
   start_time=time.time()
   x = spsolve(A.tocsr(), b)
   end_time=time.time()
   ```

   In the preceding snippet of code, we are noting down the time before the `solve` function is called and also the time immediately after the `solve` function is called.

3. Print the time taken to solve:

   ```
   print(end_time-start_time)
   0.015040397644042969
   ```

The output suggests that the time taken to solve the equation was `0.01` seconds when working using the sparse matrix format.

Now that we understand that the sparse matrix takes `0.01` seconds to solve the equation, we will now consider the time taken to compute if it were not in sparse matrix format.

In order to do that, we will use a similar procedure to what we did with the sparse matrix, but on a regular dense matrix this time. So, we will be using the `solve` function in place of the `spsolve` function:

```
import time
start_time=time.time()
x = solve(A.toarray(), b)
end_time=time.time()
print(end_time-start_time)
12.89
```

We have converted the linked list matrix into a regular matrix by using the `.toarray()` function on top of it.

Also, we can now notice the huge difference in the time taken to solve the equation using the sparse matrix format, which was `0.01` seconds, and the dense (regular) matrix, which was `12.89` seconds.

Thus, sparse matrices not only help in the storage of an array, but also in speeding up the computations quite a bit.

6
Solving Equations and Optimization

In this chapter, you will learn how to solve a system of equations and its methods of solution, as well as the optimization solutions for non-linear equations.

We will see the following topics:

- Non-linear equations and systems
- System of equations and how to solve it
- Choosing the solver used to find the solution of equations
- Solving constrained non-linear optimization problems in several variables
- Solving one-dimensional optimization problems
- Solving multidimensional non-linear equations using the Newton-Krylov method
- Solving multidimensional non-linear equations using the Anderson method
- Finding the best linear fit for a set of data
- Doing non-linear regression for a set of data
- Regression

Introduction

One way to measure the success or failure of the real equation is by computing the uniform norm of the difference between the original function and the interpolation. In this particular case, that norm is close to 2.0. We could approximate this value by performing the following computation on a large set of points in the domain:

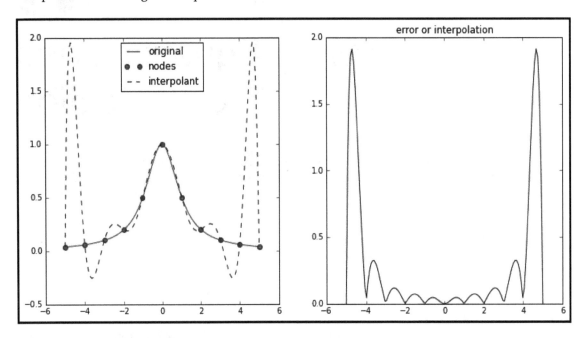

However, this is a crude approximation of the actual error. To compute the true norm, we need a mechanism that calculates the actual maximum value of a function over a finite interval, rather than over a discrete set of points. To perform this operation for the current example, we will use the `minimize_scalar` routine from the `scipy.optimize` module.

Non-linear equations and systems

In the solution of linear equations and systems, $f(x) = 0$, we had the choice of using either direct methods or iterative processes. A direct method in that setting was simply the application of an exact formula involving only the four basic operations: addition, subtraction, multiplication, and division. The issues with this method arise when cancellation occurs, mainly whenever sums and subtractions are present. Iterative methods, rather than computing a solution in a finite number of operations, calculate closer and closer approximations to the said solution, improving the accuracy with each step.

In the case of non-linear equations, direct methods are seldom a good idea. Even when a formula is available, the presence of non-basic operations leads to uncomfortable rounding errors. Let's see this using a very basic example.

Getting ready

Consider the quadratic equation $ax^2 + bx + c = 0$, with $a = 10^{-10}$, $b = -(10^{10} + 1)/10^{10}$, and $c = 1$. These are the coefficients of the expanded version of the polynomial $p(x) = 10^{-10}(x-1)(x-10^{10})$, with the obvious roots $x = 1$ and $x = 10^{10}$. Notice the behavior of the quadratic formula in the following command:

```
In [1]: import numpy as np
In [2]: a, b, c = 1.0e-10, -(1.0e10 + 1.)/1.0e10, 1.
In [3]: (-b - np.sqrt(b**2 - 4*a*c))/(2*a)
Out[3]: 1.00000000082740371
```

How to do it...

A notable rounding error due to cancellation has spread. It is possible to fix the situation, in this case, by multiplying the numerator and denominator of this formula by the conjugate of its denominator and using the resulting formula instead:

```
In [4]: 2*c / (-b + np.sqrt(b**2 - 4*a*c))
Out[4]: 1.0
```

How it works...

Even the algebraic solvers coded in the SymPy libraries share this defect, as the following example shows:

 The SymPy libraries have a set of algebraic solvers and all of them are accessed from the common `solve` routine. Currently, this method solves univariate polynomials, transcendental equations, and a piecewise combination of them. It also solves systems of linear and polynomial equations. For more information, refer to the official documentation for SymPy at http://docs.sympy.org/dev/modules/solvers/solvers.html.

```
In [5]: from sympy import symbols, solve
In [6]: x = symbols('x', real=True)
In [7]: solve(a*x**2 + b*x + c)
Out[7]: [1.00000000000000, 9999999999.00000]
```

To avoid having to second-guess the accuracy of our solutions or fine-tune each possible formula that solves a non-linear equation, we can always adopt iterative processes to achieve arbitrarily close approximations.

System of equations and how to solve it

In this recipe, we will see how to solve system of equations and its applications.

Getting ready

Get the library and the `numpy.linalg.solve` reference.

How to do it...

We have the following functions and parameters:

Parameters	`a : (..., M, M) array_like`
	Coefficient matrix.
	`b : {(..., M,), (..., M, K)}, array_like`
	Ordinate or dependent variable values.

Returns	x : {(..., M,), (..., M, K)} ndarray Solution to the system a x = b. The returned shape is identical to b.
Raises	LinAlgError If a is singular or not square.

Broadcasting rules apply; see the numpy.linalg documentation for details. The solutions are computed using the LAPACK _gesv routine it must be square and of full-rank, that is, all rows (or, equivalently, columns) must be linearly independent; if either is not true, use lstsq for the least-squares best *solution* of the system/equation.

How it works...

Solve the system of equations $3 * x_0 + x_1 = 9$ and $x_0 + 2 * x_1 = 8$:

```
a = np.array([[3,1], [1,2]])
b = np.array([9,8])
x = np.linalg.solve(a, b)
x
array([ 2.,   3.])
```

Check if the solution is correct:

```
np.allclose(np.dot(a, x), b)
True
```

Choosing the solver used to find the solution of equations

How about having the functions for computing equations and finding the solutions? In this section, we will see how to do it.

Getting ready

This is a collection of general-purpose, non-linear multidimensional solvers. These solvers find x for which $F(x) = 0$. Both x and F can be multidimensional.

Functions	Parameters
`newton_krylov(F, xin[, iter, rdiff, method, ...])`	Find a root of a function, using Krylov approximation for inverse Jacobian.
`anderson(F, xin[, iter, alpha, w0, M, ...])`	Find a root of a function, using (extended) Anderson mixing.

General non-linear solvers:

Functions	Parameters
`broyden1(F, xin[, iter, alpha, ...])`	Find a root of a function, using Broyden's first Jacobian approximation.
`broyden2(F, xin[, iter, alpha, ...])`	Find a root of a function, using Broyden's second Jacobian approximation.

Simple iterations:

Fucntions	Parameters
`excitingmixing(F, xin[, iter, alpha, ...])`	Find a root of a function, using a tuned diagonal Jacobian approximation.
`linearmixing(F, xin[, iter, alpha, verbose, ...])`	Find a root of a function, using a scalar Jacobian approximation.
`diagbroyden(F, xin[, iter, alpha, verbose, ...])`	Find a root of a function, using diagonal Broyden Jacobian approximation.

How to do it...

We can use the following code, which works to have the solution of the equation:

```
def F(x):
...     return np.cos(x) + x[::-1] - [1, 2, 3, 4]
import scipy.optimize
x = scipy.optimize.broyden1(F, [1,1,1,1], f_tol=1e-14)
x
array([ 4.04674914,  3.91158389,  2.71791677,  1.61756251])
np.cos(x) + x[::-1]
array([ 1.,   2.,   3.,   4.])
```

How it works...

This is an example of `newton_krylov`:

```
import numpy as np
from scipy.optimize import newton_krylov
from numpy import cosh, zeros_like, mgrid, zeros

# parameters
nx, ny = 75, 75
hx, hy = 1./(nx-1), 1./(ny-1)

P_left, P_right = 0, 0
P_top, P_bottom = 1, 0

def residual(P):
    d2x = zeros_like(P)
    d2y = zeros_like(P)

    d2x[1:-1] = (P[2:]    - 2*P[1:-1] + P[:-2]) / hx/hx
    d2x[0]    = (P[1]     - 2*P[0]    + P_left)/hx/hx
    d2x[-1]   = (P_right - 2*P[-1]    + P[-2])/hx/hx

    d2y[:,1:-1] = (P[:,2:] - 2*P[:,1:-1] + P[:,:-2])/hy/hy
    d2y[:,0]    = (P[:,1]   - 2*P[:,0]    + P_bottom)/hy/hy
    d2y[:,-1]   = (P_top    - 2*P[:,-1]   + P[:,-2])/hy/hy

    return d2x + d2y - 10*cosh(P).mean()**2

# solve
guess = zeros((nx, ny), float)
sol = newton_krylov(residual, guess, method='lgmres', verbose=1)
print('Residual: %g' % abs(residual(sol)).max())
```

```
# visualize
import matplotlib.pyplot as plt
x, y = mgrid[0:1:(nx*1j), 0:1:(ny*1j)]
plt.pcolor(x, y, sol)
plt.colorbar()
plt.show()
```

In the following screenshot, we can see the result of the solution:

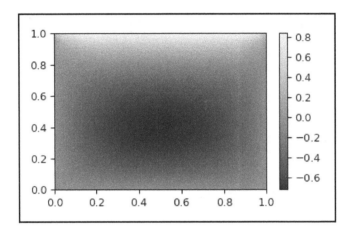

Solving constrained non-linear optimization problems in several variables

The optimization problem is best described as the search for a local maximum or minimum value of a scalar-valued function $f(x)$. This search can be performed for all possible input values in the domain of f (and, in this case, we refer to this problem as an unconstrained optimization), or for a specific subset of it that is expressible by a finite set of identities and inequalities (and we refer to this other problem as a constrained optimization). In this section, we are going to explore both modalities in several settings.

Getting ready

We focus on the search for the local minima of a function $f(x)$ in an interval $[a, b]$ (the search for local maxima can then be regarded as the search of the local minima of the function $-f(x)$ in the same interval). For this task, we have the `minimize_scalar` routine in the `scipy.optimize` module. It accepts as obligatory input a univariate function $f(x)$, together with a search method.

Most search methods are based on the idea of bracketing that we used for root finding, although the concept of a bracket is a bit different in this setting. In this case, a good bracket is a triple $x < y < z$, where $f(y)$ is less than both $f(x)$ and $f(z)$. If the function is continuous, its graph presents a U-shape on a bracket. This guarantees the existence of a minimum inside of the subinterval $[x, z]$. A successful bracketing method will look, on each successive step, for the target extremum in either $[x, y]$ or $[y, z]$.

Let's construct a simple bracketing method for testing purposes. Assume we have an initial bracket $a < c < b$. By quadratic interpolation, we construct a parabola through the points $(a, f(a))$, $(c, f(c))$, and $(b, f(b))$. Because of the U-shape condition, there must be a minimum (easily computable) for the interpolating parabola, say $(d, f(d))$. It is not hard to prove that the value d lies between the midpoints of the subintervals $[a, c]$, and $[c, b]$. We will use this point d for our next bracketing step. For example, if it happens that $c < d$, then the next bracket will be either $c < d < b$, or $a < c < d$. Easy enough! Let's implement this method.

How to do it...

This is how we represent the solution of the Lagrange equation following, this represents the solution of the system:

```
import numpy as np
from scipy.interpolate import lagrange
from scipy.optimize import OptimizeResult, minimize_scalar
def good_bracket(func, bracket):
    a, c, b = bracket
    return (func(a) > func(c)) and (func(b) > func(c))
def parabolic_step(f, args, bracket, **options):
    stop = False
    funcalls = 0
    niter = 0
    while not stop:
        niter += 1
        interpolator = lagrange(np.array(bracket), f(np.array(bracket)))
        funcalls += 3
        a, b, c = interpolator.coeffs
```

```
            d = -0.5*b/a
            if np.allclose(bracket[1], d):
                minima = d
                stop = True
            elif bracket[1] < d:
                newbracket = [bracket[1], d, bracket[2]]
                if good_bracket(f, newbracket):
                    bracket = newbracket
                else:
                    bracket = [bracket[0], bracket[1], d]
            else:
                newbracket = [d, bracket[1], bracket[2]]
                if good_bracket(f, newbracket):
                    bracket = newbracket
                else:
                    bracket = [bracket[0], d, bracket[1]]
    return OptimizeResult(fun=f(minima), x=minima, nit=niter,
nfev=funcalls)
```

The output of any minimizing method must be
an `OptimizeResult` object, with at least the x attribute (the solution to
the optimization problem). In the example we have just run, the attributes
coded in this method are x, fun (the evaluation of f at that
solution), **number of iterations (nit)**, and **number of functions
evaluations needed (nfev)**.

How it works...

Let's run this method with this example:

```
def g(x): return -np.exp(-x)*np.sin(x)
f = np.vectorize(lambda x: max(1-x, 2+x))
print (good_bracket(f, [-1, -0.5, 1]))
print (minimize_scalar(f, bracket=[-1, -0.5, 1], method=parabolic_step))

True
  fun: array(1.5000021457670878)
 nfev: 33
  nit: 11
    x: -0.50000214576708779

print (good_bracket(g, [0, 1.2, 1.5]))
True

print (minimize_scalar(g, bracket=[0,1.2,1.5], method=parabolic_step))
```

```
  fun: -0.32239694192707452
 nfev: 54
  nit: 18
    x:  0.78540558549352946
```

There are two methods already coded for univariate scalar minimization, golden, using a `golden` section search, and `brent`, following an algorithm by Brent and Dekker:

```
minimize_scalar(f, method='brent', bracket=[-1, -0.5, 1])
      fun: array(1.5)
     nfev: 22
      nit: 21
  success: True
        x: -0.5

minimize_scalar(f, method='golden', bracket=[-1, -0.5, 1])
      fun: array(1.5)
     nfev: 44
      nit: 39
  success: True
        x: -0.5

minimize_scalar(g, method='brent', bracket=[0, 1.2, 1.5])
      fun: -0.32239694194483448
     nfev: 11
      nit: 10
  success: True
        x: 0.78539816017203079

minimize_scalar(g, method='golden', bracket=[0, 1.2, 1.5])
      fun: -0.32239694194483448
     nfev: 43
      nit: 38
  success: True
        x: 0.7853981573284226
```

Solving one-dimensional optimization problems

This section represent the solution of one-dimensional optimization problems.

Getting ready

This is the function that is required to solve the method:

```
scipy.optimize.minimize(fun, x0, args=(), method=None, jac=None, hess=None,
hessp=None, bounds=None, constraints=(), tol=None, callback=None,
options=None)
```

How to do it...

Minimization of the scalar function of one or more variables:

```
minimize f(x) subject to

g_i(x) >= 0,   i = 1,...,m
h_j(x)  = 0,   j = 1,...,p
```

How it works...

This is represented in the following code:

```
from scipy.optimize import minimize, rosen, rosen_der

x0 = [1.3, 0.7, 0.8, 1.9, 1.2]
res = minimize(rosen, x0, method='Nelder-Mead', tol=1e-6)
res.x
array([ 1.,   1.,   1.,   1.,   1.])

res = minimize(rosen, x0, method='BFGS', jac=rosen_der,
...                 options={'gtol': 1e-6, 'disp': True})
Optimization terminated successfully.
        Current function value: 0.000000
        Iterations: 26
        Function evaluations: 31
        Gradient evaluations: 31
res.x
array([ 1.,   1.,   1.,   1.,   1.])
print(res.message)
Optimization terminated successfully.
res.hess_inv
array([[ 0.00749589,  0.01255155,  0.02396251,  0.04750988,  0.09495377],
# may vary
       [ 0.01255155,  0.02510441,  0.04794055,  0.09502834,  0.18996269],
       [ 0.02396251,  0.04794055,  0.09631614,  0.19092151,  0.38165151],
```

```
[ 0.04750988,  0.09502834,  0.19092151,  0.38341252,  0.7664427 ],
[ 0.09495377,  0.18996269,  0.38165151,  0.7664427,   1.53713523]])
```

Then we represent the `lambda` function:

```
fun = lambda x: (x[0] - 1)**2 + (x[1] - 2.5)**2

cons = ({'type': 'ineq', 'fun': lambda x:  x[0] - 2 * x[1] + 2},
        {'type': 'ineq', 'fun': lambda x: -x[0] - 2 * x[1] + 6},
        {'type': 'ineq', 'fun': lambda x: -x[0] + 2 * x[1] + 2})

bnds = ((0, None), (0, None))

res = minimize(fun, (2, 0), method='SLSQP', bounds=bnds,
               constraints=cons)
```

Solving multidimensional non-linear equations using the Newton-Krylov method

In this section, we will cover the non-linear equations using the Newton-Krylov method.

Getting ready

The following code is required to get ready before we proceed:

```
scipy.optimize.newton_krylov(F, xin, iter=None, rdiff=None,
method='lgmres', inner_maxiter=20, inner_M=None, outer_k=10, verbose=False,
maxiter=None, f_tol=None, f_rtol=None, x_tol=None, x_rtol=None,
tol_norm=None, line_search='armijo', callback=None, **kw)
```

In the following table can see the parameters and their description of the Anderson method

Parameters	F: `function(x)` -> `f`. Function whose root to find; should take and return an array-like object.
	xin: `array_like`. Initial guess for the solution.
	rdiff: `float, optional`. Relative step size to use in numerical differentiation.
	method: `{'lgmres', 'gmres', 'bicgstab', 'cgs', 'minres'}` or function. Krylov method to use to approximate the Jacobian. Can be a string, or a function implementing the same interface as the iterative solvers in `scipy.sparse.linalg`.
	The default is `scipy.sparse.linalg.lgmres`.
	inner_M: `LinearOperator` or `InverseJacobian`
	Preconditioner for the inner Krylov iteration. Note that you can also use inverse Jacobians as (adaptive) preconditioners. For example:
	`from scipy.optimize.nonlin import BroydenFirst, KrylovJacobian`
	`from scipy.optimize.nonlin import InverseJacobian`
	`jac = BroydenFirst()`
	`kjac = KrylovJacobian(inner_M=InverseJacobian(jac))`
	If the preconditioner has a method named `update`, it will be called as `update(x, f)` after each non-linear step, with x giving the current point, and f the current function value.
	`inner_tol, inner_maxiter, ...`

	Parameters to pass on to the *inner* Krylov solver. See `scipy.sparse.linalg.gmres` for details. `outer_k`: int, optional. Size of the subspace kept across LGMRES non-linear iterations. See `scipy.sparse.linalg.lgmres` documentation for details. `iter`: int, optional. Number of iterations to make. If omitted (default), make as many as required to meet tolerances. `verbose`: bool, optional. Print status to `stdout` on every iteration. `maxiter`: int, optional. Maximum number of iterations to make. If more are needed to meet convergence, `NoConvergence` is raised. `f_tol`: float, optional. Absolute tolerance (in max-norm) for the residual. If omitted, default is 6e-6. `f_rtol`: float, optional. Relative tolerance for the residual. If omitted, not used. `x_tol`: float, optional. Absolute minimum step size, as determined from the Jacobian approximation. If the step size is smaller than this, optimization is terminated as successful. If omitted, not used. `x_rtol`: float, optional. Relative minimum step size. If omitted, not used. `tol_norm`: function(vector) -> scalar, optional. Norm to use in convergence check. Default is the maximum norm. `line_search`: {None, 'armijo' (default), 'wolfe'}, optional. Which type of a line search to use to determine the step size in the direction given by the Jacobian approximation. Defaults to `armijo`. `callback`: Function, optional. Optional callback function. It is called on every iteration as `callback(x, f)` where x is the current solution and f the corresponding residual.
Returns	`sol`: ndarray. An array (of similar array type as x0) containing the final solution.
Raises	`NoConvergence`. When a solution was not found.

How to do it...

In the following code, we represent the functions to be executed according to the parameters:

```python
import numpy as np
from scipy.optimize import newton_krylov
from numpy import cosh, zeros_like, mgrid, zeros

# parameters
nx, ny = 75, 75
hx, hy = 1./(nx-1), 1./(ny-1)

P_left, P_right = 0, 0
P_top, P_bottom = 1, 0

def residual(P):
    d2x = zeros_like(P)
    d2y = zeros_like(P)

    d2x[1:-1] = (P[2:]    - 2*P[1:-1] + P[:-2]) / hx/hx
    d2x[0]    = (P[1]     - 2*P[0]    + P_left)/hx/hx
    d2x[-1]   = (P_right - 2*P[-1]   + P[-2])/hx/hx

    d2y[:,1:-1] = (P[:,2:]  - 2*P[:,1:-1] + P[:,:-2])/hy/hy
    d2y[:,0]    = (P[:,1]   - 2*P[:,0]    + P_bottom)/hy/hy
    d2y[:,-1]   = (P_top    - 2*P[:,-1]   + P[:,-2])/hy/hy

    return d2x + d2y - 10*cosh(P).mean()**2

# solve
guess = zeros((nx, ny), float)
sol = newton_krylov(residual, guess, method='lgmres', verbose=1)
print('Residual: %g' % abs(residual(sol)).max())

# visualize
import matplotlib.pyplot as plt
x, y = mgrid[0:1:(nx*1j), 0:1:(ny*1j)]
plt.pcolor(x, y, sol)
plt.colorbar()
plt.show()
```

In the following image we can represent the solution of the method:

Solving multidimensional non-linear equations using the Anderson method

In this section, we will see the multidimensional non-linear equations using the Anderson method:

Getting ready

This is the requirement that we need to see:

```
scipy.optimize.anderson(F, xin, iter=None, alpha=None, w0=0.01, M=5,
verbose=False, maxiter=None, f_tol=None, f_rtol=None, x_tol=None,
x_rtol=None, tol_norm=None, line_search='armijo', callback=None, **kw)
```

How to do it...

This table represents the function and the parameters:

Parameters	F: `function(x) -> f`. Function whose root to find; should take and return an array-like object. xin: `array_like`. Initial guess for the solution. alpha: `float`, optional. Initial guess for the Jacobian is (-1/alpha). M: `float`, optional. Number of previous vectors to retain. Defaults to 5. w0: `float`, optional. Regularization parameter for numerical stability. Compared to unity, good values of the order of 0.01. iter: `int`, optional. Number of iterations to make. If omitted (default), make as many as required to meet tolerances. verbose: `bool`, optional. Print status to `stdout` on every iteration. maxiter: `int`, optional. Maximum number of iterations to make. If more are needed to meet convergence, `NoConvergence` is raised. f_tol: `float`, optional. Absolute tolerance (in max-norm) for the residual. If omitted, default is 6e-6. f_rtol: `float`, optional. Relative tolerance for the residual. If omitted, not used.
	x_tol: `float`, optional. Absolute minimum step size, as determined from the Jacobian approximation. If the step size is smaller than this, optimization is terminated as successful. If omitted, not used. x_rtol: float, optional. Relative minimum step size. If omitted, not used. tol_norm: `function(vector) -> scalar`, optional. Norm to use in convergence check. Default is the maximum norm. line_search: `{None, 'armijo' (default), 'wolfe'}`, optional. Which type of a line search to use to determine the step size in the direction given by the Jacobian approximation. Defaults to `armijo`. callback: Function, optional. Optional callback function. It is called on every iteration as `callback(x, f)` where x is the current solution and f the corresponding residual.
Returns	sol: `ndarray`. An array (of similar array type as x0) containing the final solution.
Raises	`NoConvergence`. When a solution was not found.

How it works...

This code represents the execution of the function:

```
def F(x):
...     return np.cos(x) + x[::-1] - [1, 2, 3, 4]
import scipy.optimize
x = scipy.optimize.broyden1(F, [1,1,1,1], f_tol=1e-14)
x
array([ 4.04674914,  3.91158389,  2.71791677,  1.61756251])
np.cos(x) + x[::-1]
array([ 1.,   2.,   3.,   4.])
```

Finding the best linear fit for a set of data

In the following section, we will see the method of finding the best linear fit for a set of data.

Getting ready

This is the main requirement that we need to have:

```
numpy.polyfit(x, y, deg, rcond=None, full=False, w=None, cov=False)
```

How to do it...

This table represents the main parameters of the method:

Parameters	x: `array_like,  shape  (M,)`. *x* coordinates of the M sample points (`x[i], y[i]`). y : `array_like, shape (M,) or (M,  K)`. *y* coordinates of the sample points. Several datasets of sample points sharing the same *x* coordinates can be fitted at once by passing in a 2D array that contains one dataset per column. deg: `int`. Degree of the fitting polynomial. rcond: `float`, optional. Relative condition number of the fit. Singular values smaller than this relative to the largest singular value will be ignored. The default value is `len(x)*eps`, where `eps` is the relative precision of the float type, about 2e-16 in most cases. full: `bool`, optional. Switch determining nature of return value. When it is `False` (the default) just the coefficients are returned. When `True`, diagnostic information from the singular value decomposition is also returned. w: `array_like, shape  (M,)`, optional. Weights to apply to the *y* coordinates of the sample points. For Gaussian uncertainties, use *1/sigma (not 1/sigma**2)*. cov: `bool`, optional. Return the estimate and the covariance matrix of the estimate. If full is `True`, then `cov` is not returned.
Returns	p: `ndarray, shape (deg + 1,) or (deg + 1, K)`. Polynomial coefficients, highest power first. If *y* was 2D, the coefficients for the kth dataset would be in `p[:,k]`. `residuals, rank, singular_values, rcond`. Present only if `full = True`. Residuals of the least-squares fit, the effective rank of the scaled Vandermonde coefficient matrix, its singular values, and the specified value of rcond. For more details, see `linalg.lstsq`. V: `ndarray, shape (M,M) or (M,M,K)`. Present only if `full = False` and `cov=True`. The covariance matrix of the polynomial coefficient estimates. The diagonal of this matrix are the variance estimates for each coefficient. If y is a 2D array, then the covariance matrix for the kth dataset are in `V[:,:,k]`.
Warns	`RankWarning`

How it works ...

We have here the code for the Anderson method:

```
x = np.array([0.0, 1.0, 2.0, 3.0,  4.0,  5.0])
y = np.array([0.0, 0.8, 0.9, 0.1, -0.8, -1.0])
z = np.polyfit(x, y, 3)
z
array([ 0.08703704, -0.81349206,  1.69312169, -0.03968254])

p = np.poly1d(z)
p(0.5)
0.6143849206349179
p(3.5)
-0.34732142857143039
p(10)
22.579365079365115

p30 = np.poly1d(np.polyfit(x, y, 30))
/... RankWarning: Polyfit may be poorly conditioned...
p30(4)
-0.80000000000000204
p30(5)
-0.99999999999999445
p30(4.5)
-0.10547061179440398

import matplotlib.pyplot as plt
xp = np.linspace(-2, 6, 100)
_ = plt.plot(x, y, '.', xp, p(xp), '-', xp, p30(xp), '--')
plt.ylim(-2,2)
(-2, 2)
plt.show()
```

The result is shown in the following screenshot:

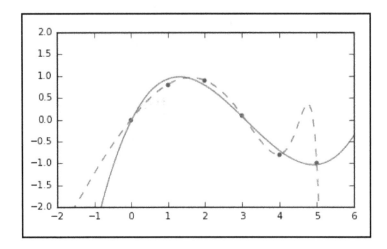

Doing non-linear regression for a set of data

In the next section, we will see how to do the non-linear regression for a set of data.

Getting ready

The following is the main requirement that we need:

```
scipy.optimize.curve_fit(f, xdata, ydata, p0=None, sigma=None,
absolute_sigma=False, check_finite=True, bounds=(-inf, inf), method=None,
jac=None, **kwargs)
```

How to do it...

We present the main functions and parameters:

The parameters are:

- `f`: Callable. The model function, `f(x, ...)`. It must take the independent variable as the first argument and the parameters to fit as separate remaining arguments.
- `xdata`: An M-length sequence or an (k,M)-shaped array for functions with k predictors. The independent variable where the data is measured.
- `ydata`: M-length sequence. The dependent data—nominally `f(xdata, ...)`.
- `p0`: None, scalar, or N-length sequence, optional. Initial guess for the parameters. If none, then the initial values will all be 1 (if the number of parameters for the function can be determined using introspection, otherwise a `ValueError` is raised).
- `sigma`: None or M-length sequence or M x M array, optional. Determines the uncertainty in `ydata`. If we define residuals as `r = ydata - f(xdata, *popt)`, then the interpretation of `sigma` depends on its number of dimensions:
- A 1D `sigma` should contain values of standard deviations of errors in `ydata`. In this case, the optimized function is `chisq = sum((r / sigma) ** 2)`. A 2D `sigma` should contain the covariance matrix of errors in `ydata`. In this case, the optimized function is `chisq = r.T @ inv(sigma) @ r`.
- New in version 0.19—None (default) is equivalent of 1D `sigma` filled with ones.
- `absolute_sigma`: `bool`, optional. If `True`, `sigma` is used in an absolute sense and the estimated parameter covariance `pcov` reflects these absolute values.
- If `False`, only the relative magnitudes of the sigma values matter. The returned parameter covariance matrix `pcov` is based on scaling sigma by a constant factor. This constant is set by demanding that the reduced chisq for the optimal parameters popt when using the scaled sigma equals unity. In other words, sigma is scaled to match the sample variance of the residuals after the fit.
- Mathematically, `pcov(absolute_sigma=False) = pcov(absolute_sigma=True) * chisq(popt)/(M-N)`.
- `check_finite`: `bool`, optional. If `True`, check that the input arrays do not contain `nans` of `infs`, and raise a `ValueError` if they do. Setting this parameter to `False` may silently produce nonsensical results if the input arrays do contain `nans`. The default is `True`.

- bounds: Two-tuple of `array_like`, optional. Lower and upper bounds on independent variables. Defaults to no bounds. Each element of the tuple must be either an array with the length equal to the number of parameters, or a scalar (in which case the bound is taken to be the same for all parameters).
  Use `np.inf` with an appropriate sign to disable bounds on all or some parameters.
- New in version 0.17—`method`: {`'lm'`, `'trf'`, `'dogbox'`}, optional. Method to use for optimization. See `least_squares` for more details. Default is `lm` for unconstrained problems and `trf` if `bounds` are provided. The method `lm` won't work when the number of observations is less than the number of variables, use `trf` or `dogbox` in this case.
- New in version 0.17—`jac` : Callable, string or None, optional. Function with signature `jac(x, ...)` which computes the Jacobian matrix of the model function with respect to parameters as a dense `array_like` structure. It will be scaled according to provided `sigma`. If None (default), the Jacobian will be estimated numerically. String keywords for the `trf` and `dogbox` methods can be used to select a finite difference scheme, see `least_squares`. New in version 0.18—`kwargs`: Keyword arguments passed
  to `leastsq`; for `method='lm'` or `least_squares` otherwise.

The returns are:

- `popt`: Array. Optimal values for the parameters so that the sum of the squared residuals of `f(xdata, *popt) - ydata` is minimized.
- `pcov`: 2D array. The estimated covariance of `popt`. The diagonals provide the variance of the parameter estimate. To compute one standard deviation errors on the parameters
  use `perr = np.sqrt(np.diag(pcov))`.
- How the `sigma` parameter affects the estimated covariance depends on the `absolute_sigma` argument, as described.
- If the Jacobian matrix of the solution doesn't have a full rank, then the `lm` method returns a matrix filled with `np.inf`. On the other hand, `trf` and `dogbox` methods use Moore-Penrose pseudoinverse to compute the covariance matrix.

The raises are:

- `ValueError`: If either `ydata` or `xdata` contain `NaN` values, or if incompatible options are used.
- `RuntimeError`: If the least-squares minimization fails.
- `OptimizeWarning`: If covariance of the parameters cannot be estimated.

How it works...

The following code represents the functions of the executed equations:

```
import numpy as np
import matplotlib.pyplot as plt
from scipy.optimize import curve_fit

def func(x, a, b, c):
...        return a * np.exp(-b * x) + c

xdata = np.linspace(0, 4, 50)
y = func(xdata, 2.5, 1.3, 0.5)
y_noise = 0.2 * np.random.normal(size=xdata.size)
ydata = y + y_noise
plt.plot(xdata, ydata, 'b-', label='data')

popt, pcov = curve_fit(func, xdata, ydata)
plt.plot(xdata, func(xdata, *popt), 'r-', label='fit')

popt, pcov = curve_fit(func, xdata, ydata, bounds=(0, [3., 2., 1.]))
plt.plot(xdata, func(xdata, *popt), 'g--', label='fit-with-bounds')

plt.xlabel('x')
plt.ylabel('y')
plt.legend()
plt.show()
```

The following is the final result:

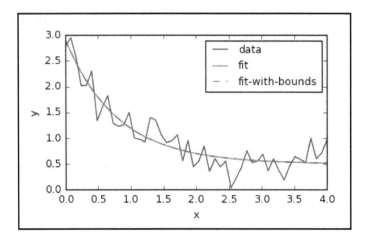

Regression

In the following section, we will see the regression and how to develop it.

Getting ready

Regression is similar to interpolation. In this case, we assume that the data is imprecise and we require an object of predetermined structure to fit the data as closely as possible. The most basic example is univariate polynomial regression to a sequence of points. We obtain that with the `polyfit` command, which we discussed briefly in the *Univariate polynomials* section of this chapter. For instance, if we want to compute the regression line in the least-squares sense for a sequence of 10 uniformly spaced points in the interval $(0, \pi/2)$ and their values under the `sin` function.

How to do it...

This code represents the main part of the function and how to solve it:

```
import numpy
import scipy
import matplotlib.pyplot as plt
x=numpy.linspace(0,1,10)
```

```
y=numpy.sin(x*numpy.pi/2)
line=numpy.polyfit(x,y,deg=1)
plt.plot(x,y,'.',x,numpy.polyval(line,x),'r')
plt.show()
```

This gives the following plot that shows linear regression with `polyfit`:

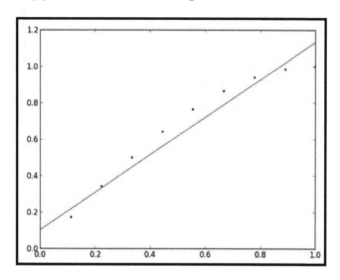

Curve fitting is also possible with splines if we use the parameters wisely. For example, in the case of univariate spline fitting that we introduced before, we can play around with the weights, smoothing factor, the degree of the smoothing spline, and so on. If we want to fit a parabolic spline for the same data as the previous example, we could issue the following commands:

```
import numpy
import scipy.interpolate
import matplotlib.pyplot as plt
x=numpy.linspace(0,1,10)
y=numpy.sin(x*numpy.pi/2)
spline=scipy.interpolate.UnivariateSpline(x,y,k=2)
xn=numpy.linspace(0,1,100)
plt.plot(x,y,'.', xn, spline(xn))
plt.show()
```

This gives a graph that shows curve fitting with splines.

How it works...

This represents the final result of the method:

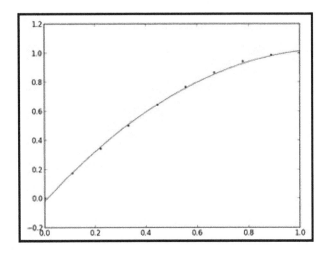

For regression from the point of view of curve fitting, there is a generic `curve_fit`
routine in the `scipy.optimize` module.

This routine minimizes the sum of squares of a set of equations using the **Levenberg-Marquardt** algorithm and offers a best fit from any kind of functions (not only polynomials
or splines). The syntax is simple:

```
curve_fit(f, xdata, ydata, p0=None, sigma=None, **kw)
```

The `f` parameter is a callable function that represents the function we seek,
and `xdata` and `ydata` are arrays of the same length that contain the x and y coordinates of
the points to be fit. The `p0` tuple holds an initial guess for the values to be found
and sigma is a vector of weights that could be used instead of the standard deviation of the
data, if necessary.

We will show its usage with a good example. We will start by generating some points on a
section of a sine wave with amplitude `A=18`, angular frequency $w=3\pi$, and phase `h=0.5`. We
corrupt the data in the `y` array; with some small random noise:

```
import numpy
import scipy
A=18; w=3*numpy.pi; h=0.5
x=numpy.linspace(0,1,100); y=A*numpy.sin(w*x+h)
y += 4*((0.5-scipy.rand(100))*numpy.exp(2*scipy.rand(100)**2))
```

We want to estimate the values of A, w, and h from the corrupted data, hence technically finding a curve fit from the set of sine waves. We start by gathering the three parameters in a list and initializing them to some values, for example, *A = 20*, *w = 2π*, and *h = 1*. We also construct a callable expression of the target function (`target_function`):

```
import scipy.optimize
p0 = [20, 2*numpy.pi, 1]
target_function = lambda x,AA,ww,hh: AA*numpy.sin(ww*x+hh)
```

We feed these, together with the fitting data, to `curve_fit` in order to find the required values:

```
pF,pVar = scipy.optimize.curve_fit(target_function, x, y, p0)
```

A sample of `pF` run on any of our experiments should give an accurate result for the three requested values:

```
print (pF)
```

The output for the preceding command is as follows:

```
[ 18.13799397 9.32232504 0.54808516]
```

This means that A was estimated to about 18.14, w was estimated very close to 3π, and h was between 0.46 and 0.55. The output of the initial data, together with a computation of the sine wave is as follows, in which original data (in blue on the left-hand side graph), corrupted (in red in both graphs), and computed sine wave (in black in the right-hand side) are shown in the following screenshot:

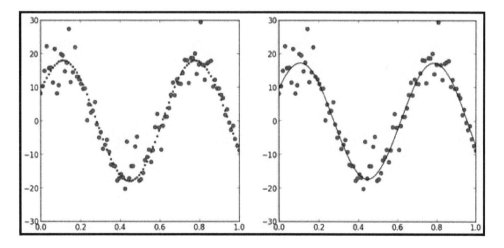

7

Constants and Special Functions

In this chapter, we will see the following recipes:

- Physical and mathematical constants available in SciPy
- Using constants in the CODATA database
- Bessel functions
- Error functions
- Orthogonal polynomials functions
- Gamma functions
- The Riemann zeta function
- Airy and Bairy functions
- The Bessel and Struve functions

Introduction

The evaluation of special functions

The `scipy.special` module contains numerically stable definitions of useful functions. Most often, the straightforward evaluation of a function at a single value is not very efficient. For instance, we would rather use a Horner scheme (`http://en.wikipedia.org/wiki/Horner%27s_method`) than a formula to find the value of a polynomial at a point. The NumPy and SciPy modules ensure that this optimization is always guaranteed with the definition of all its functions, whether by means of Horner schemes or with more advanced techniques.

Convenience and test functions

- All the convenience functions are designed to facilitate a computational environment where the user does not need to worry about relative errors. The functions seem to be pointless at first sight, but behind their codes, there are state-of-the-art ideas that offer faster and more reliable results.
- We have convenience functions beyond the ones defined in the NumPy libraries to find the solutions of trigonometric functions in degrees (`cosdg`, `sindg`, `tandg`, and `cotdg`); to compute angles in radians from their expressions in degrees, minutes, and seconds (radian); common powers *(exp2 for 2**x, and exp10 for 10**x)*; and common functions for small values of the variable (log1p for *log(1 + x)*, expm1 for *exp(x) - 1*, and cosm1 for *cos(x) - 1*).
- For instance, in the following code snippet, the log1p function computes the natural logarithm of *1 + x*. Why not simply add 1 to the value of *x* and then take the logarithm instead? Let's compare.

We need to perform the following steps:

1. Execute the following code:

```
import numpy
import scipy.special
a=scipy.special.exp10(-16)
numpy.log(1+a)
```

2. The output is as follows:

```
0.0
```

Physical and mathematical constants available in SciPy

In the following recipe, we will see how to use functions about physical and mathematical constants available in SciPy.

Getting ready...

We need to have the proper library to use the functions in SciPy installed. To install it, we need to download the library.

How to do it...

We need to see the following functions to be installed with the `scipy` library.

1. In the following table, we have mentioned some of the functions and special constants
2. It depends on the requirement you have, you need select the correct one
3. Execute the functions you have chosen

Mathematical constants:

Function	Description
pi	Pi
golden	Golden ratio
golden_ratio	Golden ratio

Physical constants:

Function	Description
speed_of_light	Speed of light in vacuum
mu_0	The magnetic constant, $\mu 0$
epsilon_0	The electric constant (vacuum permittivity), $\epsilon 0$
h	The Planck constant, h
Planck	The Planck constant, h
hbar	$\hbar = h/(2\pi)$
G	Newtonian constant of gravitation
gravitational_constant	Newtonian constant of gravitation
g	Standard acceleration of gravity
e	Elementary charge
elementary_charge	Elementary charge
R	Molar gas constant
gas_constant	Molar gas constant

`alpha`	Fine-structure constant
`fine_structure`	Fine-structure constant
`N_A`	Avogadro constant
`Avogadro`	Avogadro constant
`k`	Boltzmann constant
`Boltzmann`	Boltzmann constant
`sigma`	Stefan-Boltzmann constant, σ
`Stefan_Boltzmann`	Stefan-Boltzmann constant, σ
`Wien`	Wien displacement law constant
`Rydberg`	Rydberg constant
`m_e`	Electron mass
`electron_mass`	Electron mass
`m_p`	Proton mass
`proton_mass`	Proton mass
`m_n`	Neutron mass
`neutron_mass`	Neutron mass

Using constants in the CODATA database

In the next recipe, we will cover the constants like those used in the CODATA database.

Getting ready

The following the functions needed to be installed before we use them.

Constants database

In addition to the preceding variables of the function, `scipy.constants` also contains the 2014 CODATA recommended values database containing more physical constants.

In the following table we can see the name of the functions and the description

Function	Description
`value` (key)	Value in `physical_constants` indexed by key
`unit` (key)	Unit in `physical_constants` indexed by key
`precision` (key)	Relative precision in `physical_constants` indexed by key
`find` ([sub, disp])	Return list of `physical_constant` keys containing a given string
`ConstantWarning`	Accessing a constant no longer in the current CODATA dataset

How to do it...

The following functions represents and execute the following commands:

1. Install the proper function
2. Execute the function
3. Solve the function
4. The result will be presented

In the following table we have the name of the function and the description:

Function	Description
`c`	Speed of light in vacuum
`speed_of_light`	Speed of light in vacuum
`mu_0`	The magnetic constant, $\mu 0$
`epsilon_0`	The electric constant (vacuum permittivity), $\epsilon 0\epsilon 0$
`h`	The Planck constant, h
`Planck`	The Planck constant, h
`hbar`	$\hbar = h/(2\pi)\hbar = h/(2\pi)$
`G`	Newtonian constant of gravitation
`gravitational_constant`	Newtonian constant of gravitation
`g`	Standard acceleration of gravity

e	Elementary charge
elementary_charge	Elementary charge
R	Molar gas constant
gas_constant	Molar gas constant
alpha	Fine-structure constant
fine_structure	Fine-structure constant
N_A	Avogadro constant
Avogadro	Avogadro constant
k	Boltzmann constant
Boltzmann	Boltzmann constant
sigma	Stefan-Boltzmann constant, σ
Stefan_Boltzmann	Stefan-Boltzmann constant, σ
Wien	Wien displacement law constant
Rydberg	Rydberg constant
m_e	Electron mass
electron_mass	Electron mass
m_p	Proton mass
proton_mass	Proton mass
m_n	Neutron mass
neutron_mass	Neutron mass

Bessel functions

In this section, we will explore the fundamentals of Bessel functions and their applications.

Getting ready...

The requirement is to have the SymPy library installed to use the following functions.

How to do it...

Take the following steps to use a Bessel function:

1. Install the proper function
2. Execute the function
3. Solve the function
4. The result will be presented

In the following table we show in the first column the name of the function and the description of each one.

Function	Description
`jv(v, z)`	Bessel function of the first kind of real order and complex argument
`jn(v, z)`	Bessel function of the first kind of real order and complex argument
`jve(v, z)`	Exponentially scaled Bessel function of order v
`yn(n, x)`	Bessel function of the second kind of integer order and real argument
`yv(v, z)`	Bessel function of the second kind of real order and complex argument
`yve(v, z)`	Exponentially scaled Bessel function of the second kind of real order
`kn(n, x)`	Modified Bessel function of the second kind of integer order n
`kv(v, z)`	Modified Bessel function of the second kind of real order v
`kve(v, z)`	Exponentially scaled modified Bessel function of the second kind
`iv(v, z)`	Modified Bessel function of the first kind of real order
`ive(v, z)`	Exponentially scaled modified Bessel function of the first kind
`hankel1(v, z)`	Hankel function of the first kind
`hankel1e(v, z)`	Exponentially scaled Hankel function of the first kind
`hankel2(v, z)`	Hankel function of the second kind
`hankel2e(v, z)`	Exponentially scaled Hankel function of the second kind

Error functions

In the following section, we will learn about the error functions.

Getting ready...

Install the proper library, according to the function used.

How to do it...

Error functions are used as follows:

1. Install the proper function
2. Execute the function
3. Solve the function
4. The result will be presented

Here we have the functions and their description:

Function	Description
erf(z)	Returns the error function of complex argument
erfc(x)	Complementary error function, *1 - erf(x)*
erfcx(x)	Scaled complementary error function, *exp(x**2) * erfc(x)*
erfi(z)	Imaginary error function, *-i erf(i z)*
erfinv(y)	Inverse function for erf
erfcinv(y)	Inverse function for erfc
wofz(z)	Faddeeva function
dawsn(x)	Dawson's integral
fresnel(z)	Fresnel sin and cos integrals
fresnel_zeros(nt)	Compute nt complex zeros of sine and cosine Fresnel integrals S(z) and C(z)
modfresnelp(x)	Modified Fresnel positive integrals

modfresnelm(x)	Modified Fresnel negative integrals

The following are not universal functions.

erf_zeros(nt)	Compute nt complex zeros of error function erf(z)
fresnelc_zeros(nt)	Compute nt complex zeros of cosine Fresnel integral C(z)
fresnels_zeros(nt)	Compute nt complex zeros of sine Fresnel integral S(z)

Orthogonal polynomials functions

In this section, we will explore orthogonal polynomials functions and their applications.

Getting ready...

The functions in the next section evaluate values of orthogonal polynomials:

How to do it...

1. Install the proper function
2. Execute the function
3. Solve the function
4. The result will be presented

Here we can see the functions and the description.

Function	Description
assoc_laguerre(x, n[, k])	Compute the generalized (associated) Laguerre polynomial of degree n and order k
eval_legendre(n, x[, out])	Evaluate Legendre polynomial at a point
eval_chebyt(n, x[, out])	Evaluate Chebyshev polynomial of the first kind at a point

`eval_chebyu(n, x[, out])`	Evaluate Chebyshev polynomial of the second kind at a point
`eval_chebyc(n, x[, out])`	Evaluate Chebyshev polynomial of the first kind on [-2, 2] at a point
`eval_chebys(n, x[, out])`	Evaluate Chebyshev polynomial of the second kind on [-2, 2] at a point
`eval_jacobi(n, alpha, beta, x[, out])`	Evaluate Jacobi polynomial at a point
`eval_laguerre(n, x[, out])`	Evaluate Laguerre polynomial at a point
`eval_genlaguerre(n, alpha, x[, out])`	Evaluate generalized Laguerre polynomial at a point
`eval_hermite(n, x[, out])`	Evaluate physicist's Hermite polynomial at a point
`eval_hermitenorm(n, x[, out])`	Evaluate probabilist's (normalized) Hermite polynomial at a point
`eval_gegenbauer(n, alpha, x[, out])`	Evaluate Gegenbauer polynomial at a point
`eval_sh_legendre(n, x[, out])`	Evaluate shifted Legendre polynomial at a point
`eval_sh_chebyt(n, x[, out])`	Evaluate shifted Chebyshev polynomial of the first kind at a point
`eval_sh_chebyu(n, x[, out])`	Evaluate shifted Chebyshev polynomial of the second kind at a point
`eval_sh_jacobi(n, p, q, x[, out])`	Evaluate shifted Jacobi polynomial at a point

The following functions compute roots and quadrature weights for orthogonal polynomials:

Function	Description
`roots_legendre(n[, mu])`	Gauss-Legendre quadrature
`roots_chebyt(n[, mu])`	Gauss-Chebyshev (first kind) quadrature
`roots_chebyu(n[, mu])`	Gauss-Chebyshev (second kind) quadrature

`roots_chebyc(n[, mu])`	Gauss-Chebyshev (first kind) quadrature
`roots_chebys(n[, mu])`	Gauss-Chebyshev (second kind) quadrature
`roots_jacobi(n, alpha, beta[, mu])`	Gauss-Jacobi quadrature
`roots_laguerre(n[, mu])`	Gauss-Laguerre quadrature
`roots_genlaguerre(n, alpha[, mu])`	Gauss-generalized Laguerre quadrature
`roots_hermite(n[, mu])`	Gauss-Hermite (physicist's) quadrature
`roots_hermitenorm(n[, mu])`	Gauss-Hermite (statistician's) quadrature
`roots_gegenbauer(n, alpha[, mu])`	Gauss-Gegenbauer quadrature
`roots_sh_legendre(n[, mu])`	Gauss-Legendre (shifted) quadrature
`roots_sh_chebyt(n[, mu])`	Gauss-Chebyshev (first kind, shifted) quadrature
`roots_sh_chebyu(n[, mu])`	Gauss-Chebyshev (second kind, shifted) quadrature
`roots_sh_jacobi(n, p1, q1[, mu])`	Gauss-Jacobi (shifted) quadrature

Gamma function

In the next section, we will cover the gamma function.

Getting ready...

To use and execute the functions, we need to have installed the proper library included in the package.

How to do it...

Steps to use gamma function are as follows:

1. Install the proper function
2. Execute the function
3. Solve the function
4. The result will be presented

We will see in the following table the function and the description:

Function	Description
`gamma(z)`	Gamma function
`gammaln(x, /[, out, where, casting, order, ...])`	Logarithm of the absolute value of the gamma function
`loggamma(z[, out])`	Principal branch of the logarithm of the gamma function
`gammasgn(x)`	Sign of the gamma function
`gammainc(a, x)`	Regularized lower incomplete gamma function
`gammaincinv(a, y)`	Inverse to `gammainc`
`gammaincc(a, x)`	Regularized upper incomplete gamma function
`gammainccinv(a, y)`	Inverse to `gammaincc`
`beta(a, b)`	Beta function
`betaln(a, b)`	Natural logarithm of absolute value of beta function
`betainc(a, b, x)`	Incomplete beta integral
`betaincinv(a, b, y)`	Inverse function to beta integral
`psi(z[, out])`	The digamma function
`rgamma(z)`	Gamma function inverted
`polygamma(n, x)`	Polygamma function n
`multigammaln(a, d)`	Returns the log of multivariate gamma, also sometimes called the **generalized gamma**
`digamma(z[, out])`	The digamma function
`poch(z, m)`	Rising `factorial (z)_m`

How it works...

The working of gamma function is explained as follows:

- The gamma function is a logarithmic, convex, smooth function operating on complex numbers, which interpolates the factorial function for all non-negative integers. It is not defined at zero or any negative integer. This is the most common special function and is widely used in many different applications, either by itself or as the main ingredient in the definition of many other functions. The gamma function is used in diverse fields such as quantum physics, astrophysics, statistics, and fluid dynamics.
- The gamma function is defined by the improper integral, as follows:

$$\Gamma(z) = \int_0^\infty e^{-t} t^{z-1} \, dt$$

- Evaluation of gamma at integer values gives shifted factorials, and that is precisely how the factorials are coded in SciPy.
- The `scipy.special` module has algorithms to obtain a fast evaluation of the gamma function at any permissible value. It also contains routines for performing evaluation of the most common compositions of the gamma functions appearing in the literature: `gammaln` for the natural logarithm of the absolute value of gamma, `rgamma` for the value one over gamma, beta for quotients, and `betaln` for the natural logarithm of the latter. We also have implementations of the logarithm of its derivative (`psi`).
- An obvious application of gamma functions is the ability to perform computations that are virtually impossible for a computer if approached in a direct way. For instance, in statistical applications we often work with ratios of factorials. If these factorials are too large for the precision of a computer, we resort to expressions involving their logarithms instead. Even then, computing *ln(a! / b!)* can prove to be an impossible task (try, for example, with a = 10**15 and b = a - 10**10). An elegant solution uses the digamma `psi` function by an application of the mean value theorem on the *ln(gamma(x))* function. With proper estimation, we obtain the excellent approximation (for this choice of *a* and *b*):

$$ln(a!/b!) \simeq 10^{10} \psi(a)$$

Let's take a look at the following code snippet:

```
import scipy.special
10**10*scipy.special.psi(10**15)
```

The output is as follows:

```
345387763949.10681
```

The Riemann zeta function

In the next section, we will see the Riemann zeta function.

Getting ready

We need to have installed the library according to this function in the package.

The Riemann zeta function is very important in analytic number theory and has applications in physics and the probability theory. It computes the *p*-series for any complex *p* value, *this is the final function*:

$$\zeta(p) = \sum_{n=1}^{\infty} \frac{1}{n^p}$$

How to do it...

The definition coded in SciPy allows a more flexible generalization of this function, as follows:

1. Execute the following with the function in step 2
2. Solve the equation of the function with the proper function accordingly

$$\zeta(a, p) = \sum_{n=0}^{\infty} \frac{1}{(n+a)^p}$$

How it works...

Among others, this function has applications in the field of particle physics and in dynamical systems (http://en.wikipedia.org/wiki/Hurwitz_zeta_function).

Airy and Bairy functions

In the next section, we will see Airy and Bairy functions.

Getting ready...

These are solutions of the Stokes equation and are obtained by solving the following differential equation:

$$y'' = xy$$

How to do it...

Airy and Bairy functions are used as follows:

1. This equation has two linearly independent solutions, both of them defined as an improper integral for real values of the independent variable.
2. The `airy` command computes both functions (`Ai` and `Bi`) as well as their corresponding derivatives (`Aip` and `Bip`, respectively).
3. In the following code, we take advantage of the `contourf` command in `matplotlib.pyplot` to present an image of the real part of the output of the Bairy function `Bi` for an array of 801 x 801 complex values uniformly spaced in the square from *-4 - 4j* to *4 + 4j*.
4. We also offer this graph as a surface plot using the `mplot3d` module of `mpl_toolkits` as follows:

```
import numpy
import scipy.special
import matplotlib.pyplot as plt
import mpl_toolkits.mplot3d
x=numpy.mgrid[-4:4:100j,-4:4:100j]
z=x[0]+1j*x[1]
(Ai, Aip, Bi, Bip) = scipy.special.airy(z)
```

```
steps = range(int(Bi.real.min()), int(Bi.real.max()),6)
fig=plt.figure()
subplot1=fig.add_subplot(121,aspect='equal')
subplot1.contourf(x[0], x[1], Bi.real, steps)
```

The Bessel and Struve functions

In the next section, we will get into the details of the Bessel and Struve functions, and I am sure you are going to enjoy learning about them.

Getting ready...

Bessel functions are both of the canonical solutions to Bessel's homogeneous differential equation:

$$x^2 y'' + xy' + (x^2 = a^2)y = 0$$

How to do it...

The Bessel and Struve functions are used as follows:

1. These equations arise naturally in the solution of Laplace's equation in cylindrical coordinates
2. The solutions of the non-homogeneous Bessel differential equation shown in the following diagram are called **Struve** functions
3. We will follow the next equation and apply the proper function accordingly:

$$x^2 y'' + xy' + (x^2 = a^2)y = \frac{4(x/2)^{a+1}}{\sqrt{\pi}(a + \frac{1}{2})}$$

How it works...

These functions work as follows:

1. In either case, the order of the equation is the complex number alpha which acts as a parameter. Depending on the canonical solution and the order, the Bessel and Struve functions are addressed (and computed) differently.

2. For Bessel functions, we have algorithms to produce Bessel functions of the first kind (`jv`) and second the kind (`yn` and `yv`), Hankel functions of the first and second kind (`hankel1` and `hankel2`), and the modified Bessel functions of the first and second kind (`iv`, `kn`, and `kv`). Their syntax is similar in all cases: the first parameter is the order and the second parameter the independent variable. The *n* component in the definition indicates that an integer is to be used as the order (since they are optimally coded for that situation).

There's more

Other special functions

There are more special functions included in the `scipy.special` module that are of great use in many applications in both pure and applied mathematics. An exhaustive list would be too large for the scope of this chapter and I encourage you to use the different utilities for each set of special functions. Among the most interesting ones, we have **elliptic functions**, **Gauss hypergeometric functions**, **parabolic cylinder functions**, **Mathieu functions**, **spheroidal wave functions**, and **Kelvin functions**.

8

Calculus, Interpolation, and Differential Equations

In this chapter, we will present the following recipes:

- Integration
- Computing integrals using a Gaussian quadrature
- Computing integrals with weighting functions
- Computing multiple integrals
- Interpolation
- Computing a polynomial interpolation for a set of data points
- Univariate interpolation
- Finding a cubic spline that interpolates a set of data
- Defining a B-spline for a given set of control points
- Differentiation
- Solving a one-dimensional ordinary differential equation
- Solving a system of ordinary differential equations
- Solving differential equations and systems with parameters
- Using ode and the objected-oriented interface to solve differential equations

Introduction

Common to the design of a railway or road building (especially for highway exits), as well as those crazy loops in many roller coasters, is the solution of differential equations in two or three dimensions that address the effect of curvature and centripetal acceleration on moving bodies. In the 1970s, Werner Stengel studied and applied several models to attack this problem and, among the many solutions he found, one struck him as particularly brilliant—the employment of clothoid loops (based on sections of Cornu's spiral). The first looping coaster designed with this paradigm was constructed in 1976 in the Baja Ridge area of Six Flags Magic Mountain, in Valencia, California, USA. It was built during the great American Revolution, and it featured the very first vertical loop (together with two corkscrews, for a total of three inversions). The following is an image of a roller coaster for you to relate this to:

The tricky part of the design was based on a system of differential equations, whose solution depended on the integration of Fresnel-type sine and cosine integrals, and then selecting the appropriate sections of the resulting curve. Let's see the computation and plot of these interesting functions:

```python
import numpy as np
from scipy.special import fresnel
import pylab
t = np.linspace(-10, 10, 1000)
pylab.plot(*fresnel(t), c='k')
pylab.show()
```

This results in the following plot:

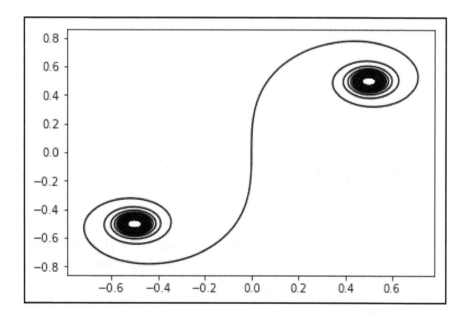

The importance of Fresnel integrals granted them a permanent place in SciPy libraries. There are many other useful integrals that shared the same fate and now lie ready for action in the `scipy.special` module. For a complete list of all of those integrals, as well as the implementation of other relevant functions and their roots or derivatives, refer to the online documentation of `scipy.special` at: `http://docs.scipy.org/doc/scipy-0.13.0/reference/special.html`.

Integration

In the next section, we will get ourselves acquainted with the methods of integration.

Getting ready

To get involved in the following recipe, we need to know about the following instructions and requirements:

- To achieve the definite integration of functions on suitable domains, we have mainly two methods—**numerical integration** and **symbolic integration**.
- Numerical integration refers to the approximation of a definite integral via a quadrature process. Depending on how the function $f(x)$ is given, the domain of integration, the knowledge of its singularities, and the choice of quadrature is the main part of this section.
- In many cases, it is also possible to perform exact integration, even for non-bounded domains, with the aid of symbolic computation. In the SciPy stack, to this effect, we have an implementation of the Risch algorithm for elementary functions, and Meijer G-functions for non-elementary integrals. Both methods are housed in the SymPy libraries. Unfortunately, these symbolic procedures do not work for all functions, and due to the complexity of the generated codes in general, the solutions obtained by this method are by no means as fast as any numerical approximation.

How to do it...

Symbolic integration is done through the following steps:

1. The definite integral of a polynomial function on a finite domain [a,b] can be computed very accurately via the fundamental theorem of calculus, using the `numpy.polynomial` module. For instance, to calculate the integral of the polynomial *p(x)=x5* on the interval [-1,1].
2. We could issue the following part integrating both parts.
3. You need to do the following:

```
In [1]: import numpy as np
In [2]: p = np.poly1d([1,0,0,0,0,0]); \
   ...: print (p)
   ...: print (p.integ())
   5
1 x
       6
0.1667 x
In [3]: p.integ()(1.0) - p.integ()(-1.0)
Out[3]: 0.0
```

5. In general, obtaining exact values for a definite integral of a generic function is hard and computationally inefficient. This is possible in some cases through symbolic integration with the aid of the Risch algorithm (for elementary functions) and Meijer G-functions (for non-elementary integrals). Both methods can be called with the common routine, `integrate`, in the SymPy library. The routine is clever enough to decide which algorithm to use, depending on the source function.
6. Let's show you the following example starting with the definite integral of the polynomial from the previous case:

```
In [4]: from sympy import integrate, symbols
In [5]: x, y = symbols('x y', real=True)
In [6]: integrate(x**5, x)
Out[6]: x**6/6
In [7]: integrate(x**5, (x, -1, 1))
Out[7]: 0
```

How it works...

Let's try something more complicated—the definite integral of the $f(x) = e^{-x}sinx$ function on the [0,1] interval:

```
In [8]: from sympy import N, exp as Exp, sin as Sin
In [9]: integrate(Exp(-x) * Sin(x), x)
Out[9]: -exp(-x)*sin(x)/2 - exp(-x)*cos(x)/2
In [10]: integrate(Exp(-x) * Sin(x), (x, 0, 1))
Out[10]: -exp(-1)*sin(1)/2 - exp(-1)*cos(1)/2 + 1/2
In [11]: N(_)
Out[11]: 0.245837007000237
```

Symbolic integration, when it works, treats singularities the right way, as follows:

```
In [12]: integrate(Sin(x) / x, x)
Out[12]: Si(x)
In [13]: integrate(Sin(x) / x, (x, 0, 1))
Out[13]: Si(1)
In [14]: N(_)
Out[14]: 0.946083070367183
In [15]: integrate(x**1, (x, 0, 1))
Out[15]: 1/2
```

Integration over unbounded domains is also possible:

```
In [16]: from sympy import oo
In [17]: integrate(Exp(-x**2), (x,0,+oo))
Out[17]: sqrt(pi)/2
```

Computing integrals using the Newton-Cotes method

The implementation of these two algorithms does not compute the interpolators explicitly. The final formulas are the target here, and the way it is coded in SciPy is by means of Newton-Cotes quadratures.

The routines to perform Newton-Cotes are hidden (in the sense that they are not reported in the tutorials or documentation in the official pages of SciPy) and are meant to be used only internally by `cumtrapz` or `simps`. They provide only the corresponding coefficients that multiply the function evaluation at the nodes.

However, Newton-Cotes quadrature formulas are usually very accurate by themselves in the right scenarios. They can be used to compute better approximations in many cases, without needing to conform to trapezoidal or Simpson's rules.

Let's show you how it works for our running example, now with only four equally spaced nodes in the [-1,1] interval:

```
In [27]: from scipy.integrate import newton_cotes
In [28]: coefficients, abs_error = newton_cotes(3, equal=True); \
    ....: nodes = np.linspace(-1, 1, 4); \
    ....: print (coefficients)
[ 0.375 1.125 1.125 0.375]
In [29]: integral = (coefficients * f(nodes)).sum(); \
    ....: print (integral)
0.0
In [30]: from math import fsum
In [31]: integral = fsum(coefficients * f(nodes)); \
    ....: print (integral)
-7.8062556419e-18
```

If the nodes of our choice happen to be equally spaced, then there is an improvement of the trapezoidal rule in a special case—if the number of subintervals is a power of two. In that case, we may use the Romberg Rule—an improvement that uses the Richardson extrapolation. We can access it with the `romb` routine in the same module.

Let's compare the results with our running example, this time using 64 subintervals of size 1/32 in the [-1,1] interval:

```
In [32]: from scipy.integrate import romb
In [33]: nodes = np.linspace(-1, 1, 65)
In [34]: romb(f(nodes), dx=1./32)
0.0
```

We have the option to report the table that shows the Richardson extrapolation from the given nodes:

```
import scipy.integrate as integrate
import scipy.special as special
result = integrate.quad(lambda x: special.jv(2.5,x), 0, 4.5)
result
(1.1178179380783249, 7.866317248189980e-09)
```

We might not have any preference for the choice of nodes, but we still have our minds set on using Romberg's rule for our numerical integration scheme. In this case, we could use the `romberg` routine, for which we only need to provide the expression of a function and the limits of integration. Optionally, we can provide absolute or relative tolerances for the error (which are both set by default to 1.48e-8):

This code represents the proper execution of the integration:

```
from scipy.integrate import romberg
romberg(f, -1, 1, show=True)

The final result is 2.35040238729 after 33 function evaluations.
2.350402387287607
```

Computing integrals using a Gaussian quadrature

In the next sections, we will look at how to compute integrals in detail.

Getting ready

This recipe can be done by using the following integration equations.

- `quad`: General-purpose integration
- `dblquad`: General-purpose double integration
- `tplquad`: General-purpose triple integration
- `fixed_quad`: Integrate `func(x)` using Gaussian quadrature of order n
- `quadrature`: Integrate with the given tolerance using Gaussian quadrature
- `romberg`: Integrate `func` using Romberg integration

How to do it...

There are other possibilities is to use Gaussian quadrature formulas, they are:

- These are more powerful, since the accuracy of the approximations is gained through internally computing the best possible choice of nodes
- There are two basic routines in the `scipy.integrate` module that perform the implementation of this algorithm: quadrature, if we want to specify tolerance, or `fixed_quad`, if we wish to specify the number of nodes (but not their locations!):

```
In [38]: from scipy.integrate import quadrature, fixed_quad
In [39]: value, absolute_error = quadrature(f, -1, 1, tol=1.49e-8);
\
    ....: print (value)
1.80904847478
In [40]: value, absolute_error = fixed_quad(f, -1, 1, n=4); \
    ....: print (value) # four nodes
1.80861639538
```

- A more advanced method to perform Gaussian quadrature, using an adaptive scheme, is obtained through the `quad` function in the `scipy.integrate` module. This function is a wrapper of the routine QAGS in the Fortran QUADPACK library. The algorithm breaks the domain of integration into several subintervals, and on each of them, it performs a 21-point Gauss-Kronrod quadrature rule. Further acceleration is achieved with Peter Wynn's epsilon algorithm.

For more information on QAGS as well as the other routines in the QUADPACK libraries, refer to the Netlib repositories: http://www.netlib.org/quadpack/.

How it works...

Let's see the following example:

```
In [41]: from scipy.integrate import quad
In [42]: value, absolute_error = quad(f, -1, 1); \
    ....: print (value)
1.8090484758005436
```

We can obtain the implementation details by setting the optional `full_output` argument to `True`. This gives us an additional Python dictionary with useful information:

```
In [43]: value, abs_error, info = quad(f, -1, 1, full_output=True)
In [44]: info.keys()
Out[44]: ['rlist', 'last', 'elist', 'iord', 'alist', 'blist',
          'neval']
In [45]: print ("{0} function evaluations".format(info['neval']))
21 function evaluations
In [46]: print ("Used {0} subintervals".format(info['last']))
Used 1 subintervals
```

To fully understand all of the different outputs of information, we need to know about the underlying algorithm computing the Gaussian quadratures. These particular routines use the Clensaw-Curtis method, a technique based on Chebyshev moments.

In the preceding example, by default, the code tried to use 50 Chebyshev moments. Due to the simplicity of the integrand, and since only one subinterval was needed, it was necessary only to use one of those moments. When we report the 50-entry one-dimensional outputs `rlist`, `elist`, `alist`, and `blist` from the dictionary info, we can disregard the information offered by the last 49 entries of each of them:

```
In [47]: np.set_printoptions(precision=2, suppress=True)
In [48]: print (info['rlist']) # integral approx on subintervals
[ 0.00e+000 2.32e+077 6.93e-310 0.00e+000 0.00e+000
   0.00e+000 0.00e+000 0.00e+000 0.00e+000 0.00e+000
   0.00e+000 0.00e+000 0.00e+000 0.00e+000 0.00e+000
   0.00e+000 0.00e+000 0.00e+000 0.00e+000 0.00e+000
   0.00e+000 6.45e-314 2.19e-314 6.93e-310 0.00e+000
   0.00e+000 0.00e+000 0.00e+000 0.00e+000 0.00e+000
   0.00e+000 0.00e+000 0.00e+000 0.00e+000 0.00e+000
   0.00e+000 0.00e+000 0.00e+000 0.00e+000 0.00e+000
   0.00e+000 0.00e+000 -1.48e-224 2.19e-314 6.93e-310
   0.00e+000 0.00e+000 0.00e+000 0.00e+000 0.00e+000]
In [49]: print (info['elist']) # abs error estimates on subintervals
[ 3.70e-015 2.32e+077 3.41e-322 0.00e+000 0.00e+000
   0.00e+000 0.00e+000 0.00e+000 0.00e+000 0.00e+000
   0.00e+000 0.00e+000 0.00e+000 0.00e+000 0.00e+000
   0.00e+000 0.00e+000 0.00e+000 0.00e+000 7.30e+245
   2.19e-314 6.93e-310 0.00e+000 0.00e+000 0.00e+000
   4.74e+246 2.20e-314 6.93e-310 0.00e+000 0.00e+000
   0.00e+000 0.00e+000 0.00e+000 0.00e+000 -9.52e+207
   2.19e-314 6.93e-310 0.00e+000 0.00e+000 0.00e+000
   0.00e+000 0.00e+000 0.00e+000 0.00e+000 0.00e+000
   0.00e+000 2.00e+000 2.00e+000 2.27e-322 1.05e-319]
In [50]: print (info['alist']) # subintervals left end points
```

```
[-1.  2.  0.  0.  0.  0.  0.  0.  0.  0.  0.  0.  0.  0.  0.
   0.  0.  0.  0.  0.  0.  0.  0.  0.  0.  0.  0.  0.  0.  0.
   0.  0.  0.  0.  0.  0.  0.  0.  0.  0.  0.  0.  0.  0.  0.
   0.  0.  0.  0.  0.]
In [51]: print (info['blist']) # subintervals right end pts
[ 1.  2.  0.  0.  0.  0.  0.  0.  0.  0.  0.  0.  0.  0.  0.
   0.  0.  0.  0.  0.  0.  0.  0.  0.  0.  0.  0.  0.  0.  0.
   0.  0.  0.  0.  0.  0.  0.  0.  0.  0.  0.  0.  0.  0.  0.
   0.  0.  0.  2. -0.]
```

Computing integrals with weighting functions

In the following sections, we will look at how to compute integrals with weight functions.

Getting ready

The proper function needs to be executed using the equations in each case.

How to do it...

1. Weighted functions can be realized as products of the $f(x)w(x)$ kind for some smooth function $f(x)$ with a non-negative weight function $w(x)$ containing singularities.
2. An illustrative example is given by $cos(\pi x/2)/\sqrt{x}$. We could regard this case as the product of $cos(\pi x/2)$ with $w(x)=1/\sqrt{x}$. The weight presents a single singularity of $x=0$, and is smooth otherwise.
3. The usual way to treat these integrals is by means of weighted Gaussian quadrature formulas. For example, to perform principal value integrals of functions of the form $f(x)/(x-c)$, we issue quad with the `weight='cauchy'` and `wvar=c` arguments. This calls the routine QAWC from QUADPACK.

Let's experiment with the Fresnel-type sine integral of $g(x) = sin(x)/x$ on the [-1,1] interval and compare it with `romberg`:

```
In [56]: value, abs_error = quad(f, -1, 1, weight='cauchy',wvar=0);
\
   ....: print (value)
1.89216614073
In [57]: romberg(g, -1, 1)
Out[57]: 2.35040238729
```

4. In the case of integrals of functions with weights $(x-a)^\alpha(b-x)^\beta$, where a and b are the endpoints of the domain of integration and both alpha and beta are greater than -1, we issue `quad` with the `weight='alg'` and `wvar=(alpha, beta)` arguments. This calls the QAWS routine from QUADPACK.

5. Let's experiment with the Fresnel-type cosine integral of $g(x)=cos(\pi x/2)/\sqrt{x}$ on the [0,1] interval and compare it with quadrature:

```
In [58]: def f(x): return np.cos(np.pi * x * 0.5)
In [59]: def g(x): return np.cos(np.pi * x * 0.5) / np.sqrt(x)
In [60]: value, abs_error = quad(f, 0, 1, weight='alg', \
                                  wvar=(-0.5,0)); \
   ....: print (value)
1.55978680075
In [61]: quadrature(g, 0, 1)
quadrature.py:178: AccuracyWarning: maxiter (50) exceeded. Latest
difference = 3.483182e-04
Out[61]: (1.5425452942607543, 0.00034831815190772275)
```

6. If the weight has the form $w(x)=(x-a)\alpha(b-x)\beta ln(x-a)$, $w(x)=(x-a)\alpha(b-x)\beta ln(b-x)$, or $w(x)=(x-a)\alpha(b-x)\beta ln(x-a)ln(b-x)$, we issue quad with the `weight='alg-loga'`, `weight='alg-logb'`, or `weight='alg-log'` argument respectively, and in each case, `wvar=(alpha, beta)`. For example, for the function $g(x)=x2ln(x)$ on the interval [0,1], we can issue the following:

```
In [62]: def f(x): return x**2
In [63]: def g(x): return x**2 * np.log(x)
In [64]: value, abs_error = quad(f, 0, 1, weight='alg-loga', \
                                  wvar=(0,0)); \
   ....: print (value)
-0.111111111111
```

The actual value of this integral is -1/9.

Computing multiple integrals

Multiple integrals is what will interest you as we dive deep into it, through this recipe.

Getting ready

We will need to follow some instructions and install the prerequisites.

How to do it...

1. It is possible to perform multivariate numerical integration on different domains via the application of adaptive Gaussian quadrature rules
2. In the `scipy.integrate` module, we have to this effect the routines `dblquad` (double integrals), `tplquad` (triple integrals), and `nquad` (integration over multiple variables)

These routines can only compute definite integrals over type I regions:

- In two dimensions, a type I domain can be written in the form $\{(x,y) : a<x<b, f(x)<y<h(x)\}$ for two numbers a and b, and two univariate functions $f(x)$ and $h(x)$.
- In three dimensions, a type I region can be written in the form $\{(x,y,z) : a<x<b, f(x)<y<h(x), q(x,y)<z<r(x,y)\}$ for numbers a and b, univariate functions $f(x)$ and $h(x)$, and bivariate functions $q(x,y)$ and $r(x,y)$.
- In more than three dimensions, type I regions can be written sequentially in a similar manner as their double and triple counterparts. The first variable is bound by two numbers. The second variable is bounded by two univariate functions of the first variable. The third variable is bounded by two bivariate functions of the two first variables, and so on.

Let's run a numerical integration over the function of the example inline reference to the following example. Note the order in which the different variables must be introduced in the definition of the function to be integrated:

```
In [76]: def f(x, y): return np.exp(-x**2 - y**2)
In [77]: from scipy.integrate import dblquad
In [78]: dblquad(f, 0, np.inf, lambda x:0, lambda x:np.inf)
Out[78]: (0.785398163397, 6.29467149642e-09)
```

Interpolation

In this section, we will look at how interpolations can be done via the application of different methods.

Getting ready

We will need to follow some instructions and install the prerequisites.

How to do it...

1. Interpolation is a basic method in numerical computation that is obtained from a discrete set of data points, intended to find an interpolation function that represents some higher order structure that contains the data.

2. The best known example is the interpolation of a sequence of points (x_k and y_k) in a plane to obtain a curve that goes through all of the points in the order dictated by the sequence.

3. If the points in the previous sequence are in the right position and order, it is possible to find a univariate function $y = f(x)$ for which $y\_k = f(x\_k)$. It is often reasonable to request this interpolating function to be a polynomial, a rational function, or a more complex functional object. Interpolation is also possible in higher dimensions, of course. The objective of the `scipy.interpolate` module is to offer a complete set of optimally coded applications to address this problem in different settings.

4. Let's address the easiest way of interpolating data to obtain a polynomial: Lagrange interpolation. Given a sequence of different x values of size n and a sequence of arbitrary real values y of the same size n, we seek a polynomial $p(x)$ of the degree of $n - 1$ that satisfies the n constraints $p(x[k]) = y[k]$ for all k from 0 to $n - 1$. The following code illustrates how to obtain a polynomial of degree 9 that interpolates the 10 uniformly spaced values of sine in the interval (-1, 1):

```
import numpy
import matplotlib.pyplot as plt
import scipy.interpolate
x=numpy.linspace(-1,1,10); xn=numpy.linspace(-1,1,1000)
y=numpy.sin(x)
polynomial=scipy.interpolate.lagrange(x, numpy.sin(x))
plt.plot(xn,polynomial(xn),x,y,'or')
```

```
plt.show()
```

We will obtain the following plot showing the Lagrange interpolation:

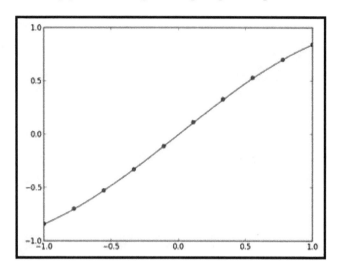

There are numerous issues with Lagrange interpolation. The first obvious drawback is that the user cannot specify the degree of the interpolation; this depends solely on the data. The procedure is also highly unstable numerically, especially for datasets with a size of over 20 points. This issue can be addressed by allowing the algorithm to depend on different properties of the dataset, rather than just the size and location of the points.

Also, it is inconvenient when we need to update the dataset by adding a few more instances; the procedure needs to be repeated from the beginning. This proves impractical if the datasets are increasing in size and are updated frequently. To address this issue, `BarycentricInterpolator` has the `add_xi` and `set_yi` methods. For example, in the next session, we start by interpolating 10 uniformly spaced values of the sine function between 1 and 10. Once done, we update the interpolating polynomial with 10 more uniformly spaced values between 1.5 and 10.5. As expected, this operation reduces the (percent) relative error of an interpolation computed at points within the interpolating ones. The following commands are used:

```
import numpy
import scipy.interpolate
x1=numpy.linspace(1,10,10); y1=numpy.sin(x1)
Polynomial=scipy.interpolate.BarycentricInterpolator(x1,y1)
exactValues=numpy.sin(x1+0.3)
exactValues
```

Here is the output for `exactValues`:

```
array([ 0.96355819, 0.74570521, -0.15774569, -0.91616594,
-0.83226744,0.0168139 , 0.85043662, 0.90217183, 0.12445442,
-0.76768581])
```

Let's find the value of `interpolatedValues` by issuing the following commands:

```
interpolatedValues=Polynomial(x1+0.3)
interpolatedValues
```

The output is as follows:

```
array([ 0.97103132, 0.74460631, -0.15742869, -0.91631362,
-0.83216445,0.01670922, 0.85059283, 0.90181323, 0.12588718,
-0.7825744 ])
```

Let's find the value of `PercentRelativeError` by issuing the following commands:

```
PercentRelativeError = numpy.abs((exactValues -
interpolatedValues)/interpolatedValues)*100
PercentRelativeError
```

The output is as follows:

```
array([ 0.76960822, 0.14758101, 0.20136334, 0.01611703, 0.01237594,
        0.62647084, 0.01836479, 0.0397652 , 1.13812858, 1.90251374])
```

Then, we find what `interpolatedValues2` holds:

```
x2=numpy.linspace(1.5,10.5,10); y2=numpy.sin(x2)
Polynomial.add_xi(x2,y2)
interpolatedValues2=Polynomial(x1+0.3)
interpolatedValues2
```

It is possible to interpolate data, not only by point location, but also with the derivatives at those locations.

The `KroghInterpolator` command allows this by including repeated *x* values and indicating the location and successive derivatives in order on the corresponding *y* values.

For instance, if we desire to construct a polynomial that is zero at the origin, one at *x* = 1, two at *x* = 2, and has horizontal tangent lines at each of these three locations, we issue the following commands:

```
import numpy
import matplotlib.pyplot as plt
import scipy.interpolate
```

```
x=numpy.array([0,0,1,1,2,2]); y=numpy.array([0,0,1,0,2,0])
interp=scipy.interpolate.KroghInterpolator(x,y)
xn=numpy.linspace(0,2,20) # evaluate polynomial in larger set
plt.plot(x,y,'o',xn,interp(xn),'r')
plt.show()
```

The result of the plot is as follows:

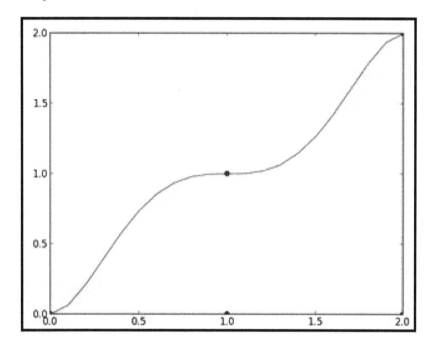

More advanced one-dimensional interpolation is possible with piecewise polynomials (`PiecewisePolynomial`). This allows control regarding the degrees of different pieces as well as the derivatives at their intersections. Other interpolation options are the `scipy.interpolate` module, **PCHIP monotonic cubic interpolation** (`pchip`), or even **univariate splines** (`InterpolatedUnivariateSpline`).

Let's examine an example with univariate splines. Its syntax is as follows:

```
InterpolatedUnivariateSpline(x, y, w=None, bbox=[None, None], k=3)
```

The x and y arrays contain dependent and independent data, respectively. The array w contains positive weights for spline fitting. The two-sequence bboxparameter specifies the boundary of the approximation interval. The last option indicates the degree of the smoothing polynomials (k).

Suppose we want to interpolate five points, as shown in the following example. These points are ordered by strictly increasing x values. We need to perform this interpolation with four cubic polynomials (one for every two consecutive points) in such a way that at least the first derivative of each two consecutive pieces agree upon their intersection. We will proceed as follows:

```
import numpy
import matplotlib.pyplot as plt
mport scipy.interpolate
x=numpy.arange(5); y=numpy.sin(x)
xn=numpy.linspace(0,4,40)
interp=scipy.interpolate.InterpolatedUnivariateSpline(x,y)
plt.plot(x,y,'.',xn,interp(xn))
plt.show()
```

This offers the following plot, showing interpolation with univariate splines:

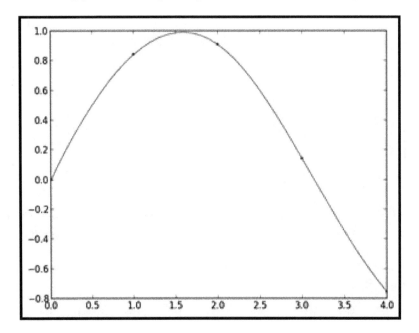

SciPy excels at interpolating in two-dimensional grids as well. It performs well with simple piecewise polynomials (LinearNDInterpolator), piecewise constants (NearestNDInterpolator), or more advanced splines (BivariateSpline). It is capable of carrying out spline interpolation on rectangular meshes in a plane (RectBivariateSpline) or on the surface of a sphere (RectSphereBivariateSpline). For unstructured data, besides the basic scipy.interpolate.BivariateSpline, it is capable of computing smooth approximations (SmoothBivariateSpline) or more involved weighted least-squares splines (LSQBivariateSpline).

The following code creates a 10 x 10 grid of uniformly spaced points in the square from (0, 0) to (9, 9), and evaluates the function $sin(x) * cos(y)$ on the points. We use these points to create a scipy.interpolate.BivariateSpline and evaluate the resulting function on the square for all values:

```
import numpy
import scipy.interpolate
import matplotlib.pyplot as plt
from mpl_toolkits.mplot3d import Axes3D
x=y=numpy.arange(10)
f=(lambda i,j: numpy.sin(i)*numpy.cos(j)) # function to interpolate
A=numpy.fromfunction(f, (10,10)) # generate samples
spline=scipy.interpolate.RectBivariateSpline(x,y,A)
fig=plt.figure()
subplot=fig.add_subplot(111,projection='3d')
xx=numpy.mgrid[0:9:100j, 0:9:100j] # larger grid for plotting
A=spline(numpy.linspace(0,9,100), numpy.linspace(0,9,100))
subplot.plot_surface(xx[0],xx[1],A)
plt.show()
```

The output is as follows, and it shows the interpolation of 2D data with bivariate splines:

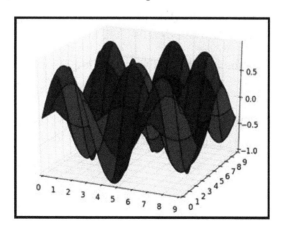

Computing a polynomial interpolation for a set of data points

In this recipe, we will look at how to compute data polynomial interpolation by applying some important methods which are discussed in detail in the coming *How to do it...* section.

Getting ready

We will need to follow some instructions and install the prerequisites.

How to do it...

Let's get started. In the following steps, we will explain how to compute a polynomial interpolation and the things we need to know:

They require the following parameters:

- `points`: An `ndarray` of floats, shape `(n, D)` data point coordinates. It can be either an array of shape `(n, D)` or a tuple of `ndim` arrays.
- `values`: An `ndarray` of float or complex shape `(n,)` data values.
- `xi`: A 2D `ndarray` of float or tuple of 1D array, shape `(M, D)`. Points at which to interpolate data.
- `method`: A `{'linear', 'nearest', 'cubic'}`—This is an optional method of interpolation.

One of the nearest return value is at the data point closest to the point of interpolation. See `NearestNDInterpolator` for more details.

- `linear` tessellates the input point set to *n*-dimensional simplices, and interpolates linearly on each simplex. See `LinearNDInterpolator` for more details.
- `cubic` (1D): Returns the value determined from a cubic spline.

- cubic (2D): Returns the value determined from a piecewise cubic, continuously differentiable (C1), and approximately curvature-minimizing polynomial surface. See `CloughTocher2DInterpolator` for more details.
- fill_value: float; optional. It is the value used to fill in for requested points outside of the convex hull of the input points. If it is not provided, then the default is nan. This option has no effect on the nearest method.
- rescale: bool; optional.

Rescale points to the unit cube before performing interpolation. This is useful if some of the input dimensions have non-commensurable units and differ by many orders of magnitude.

How it works...

One can see that the exact result is reproduced by all of the methods to some degree, but for this smooth function, the piecewise cubic interpolant gives the best results:

```
import matplotlib.pyplot as plt
import numpy as np

methods = [None, 'none', 'nearest', 'bilinear', 'bicubic', 'spline16',
           'spline36', 'hanning', 'hamming', 'hermite', 'kaiser',
'quadric',
           'catrom', 'gaussian', 'bessel', 'mitchell', 'sinc', 'lanczos']

# Fixing random state for reproducibility
np.random.seed(19680801)

grid = np.random.rand(4, 4)

fig, axes = plt.subplots(3, 6, figsize=(12, 6),
                         subplot_kw={'xticks': [], 'yticks': []})

fig.subplots_adjust(hspace=0.3, wspace=0.05)

for ax, interp_method in zip(axes.flat, methods):
    ax.imshow(grid, interpolation=interp_method, cmap='viridis')
    ax.set_title(interp_method)

plt.show()
```

This is the result of the execution:

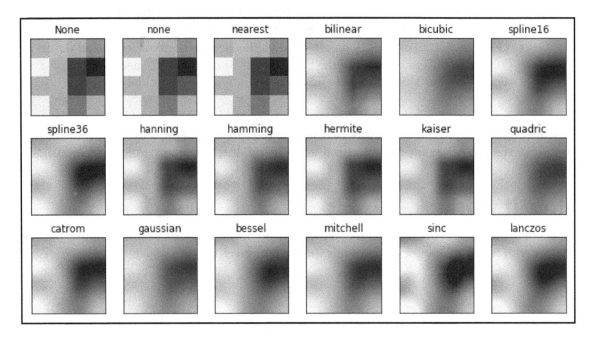

Univariate interpolation

In the next section, we will look at how to solve univariate interpolation.

Getting ready

We will need to follow some instructions and install the prerequisites.

How to do it...

The following table summarizes the different univariate interpolation modes coded in SciPy, together with the processes that we may use to resolve them:

Object-oriented implementation	Procedural implementation	Special parameters and functions
Nearest-neighbors	`interp1d(,kind='nearest')`	
Lagrange polynomial	`BarycentricInterpolator`	`barycentric_interpolate`
Hermite polynomial	`KroghInterpolator`	`krogh_interpolate`
Piecewise polynomial	`PiecewisePolynomial`	`piecewise_polynomial_interpolate`
Piecewise linear	`interp1d(,kind='linear')`	
Generic spline interpolation	`InterpolatedUnivariateSpline`	`splrep`
Zero-order spline	`interp1d(,kind='zero')`	
Linear spline	`interp1d(,kind='slinear')`	
Quadratic spline	`interp1d(,kind='quadratic')`	
Cubic spline	`interp1d(,kind='cubic')`	
PCHIP	`PchipInterpolator`	`pchip_interpolate`

Finding a cubic spline that interpolates a set of data

In this recipe, we will look at how to find a cubic spline that interpolates with the main method of spline.

Getting ready

We will need to follow some instructions and install the prerequisites.

How to do it...

We can use the following functions to solve the problems with this parameter:

- x: `array_like`, `shape (n,)`. A 1D array containing values of the independent variable. The values must be real, finite, and in strictly increasing order.
- y: `array_like`. An array containing values of the dependent variable. It can have an arbitrary number of dimensions, but the length along `axis` must match the length of x. The values must be finite.
- `axis`: `int`; optional. The axis along which y is assumed to be varying, meaning for x[i], the corresponding values are `np.take(y, i, axis=axis)`. The default is 0.
- `bc_type`: String or two-tuple; optional. Boundary condition type. Two additional equations, given by the boundary conditions, are required to determine all coefficients of polynomials on each segment. Refer to: `https://docs.scipy.org/doc/scipy-0.19.1/reference/generated/scipy.interpolate.CubicSpline.html#r59`.

If `bc_type` is a string, then the specified condition will be applied at both ends of a spline. The available conditions are:

- `not-a-knot` (default): The first and second segment at a curve end are the same polynomial. This is a good default when there is no information about boundary conditions.

- periodic: The interpolated function is assumed to be periodic in the period x[-1] – x[0]. The first and last value of y must be identical: y[0] == y[-1]. This boundary condition will result in y'[0] == y'[-1] and y''[0] == y''[-1].
- clamped: The first derivatives at the curve ends are zero. Assuming there is a 1D y, bc_type=((1, 0.0), (1, 0.0)) is the same condition.
- natural: The second derivatives at the curve ends are zero. Assuming there is a 1D y, bc_type=((2, 0.0), (2, 0.0)) is the same condition.

If bc_type is two-tuple, the first and the second value will be applied at the curve's start and end respectively. The tuple value can be one of the previously mentioned strings (except periodic) or a tuple (order, deriv_values), allowing us to specify arbitrary derivatives at curve ends:

- order: The derivative order; it is 1 or 2.
- deriv_value: An array_like containing derivative values. The shape must be the same as y, excluding the axis dimension. For example, if y is 1D, then deriv_value must be a scalar. If y is 3D with shape (n0, n1, n2) and axis=2, then deriv_value must be 2D and have the shape (n0, n1).
- extrapolate: {bool, 'periodic', None}; optional.

bool, determines whether or not to extrapolate to out-of-bounds points based on first and last intervals, or to return NaNs. periodic, periodic extrapolation is used. If none (default), extrapolate is set to periodic for bc_type='periodic' and to True otherwise.

How it works...

We have the following example:

```
%pylab inline
from scipy.interpolate import CubicSpline
import matplotlib.pyplot as plt
x = np.arange(10)
y = np.sin(x)
cs = CubicSpline(x, y)
xs = np.arange(-0.5, 9.6, 0.1)
plt.figure(figsize=(6.5, 4))
plt.plot(x, y, 'o', label='data')
plt.plot(xs, np.sin(xs), label='true')
plt.plot(xs, cs(xs), label="S")
plt.plot(xs, cs(xs, 1), label="S'")
```

```
plt.plot(xs, cs(xs, 2), label="S''")
plt.plot(xs, cs(xs, 3), label="S'''")
plt.xlim(-0.5, 9.5)
plt.legend(loc='lower left', ncol=2)
plt.show()
```

We can see the result here:

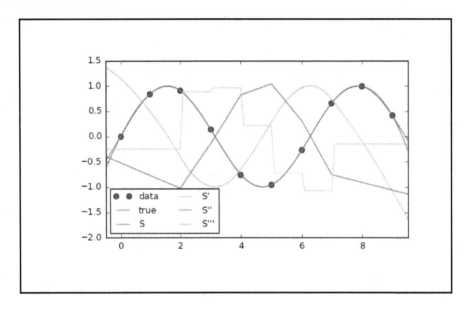

We see the next example:

```
theta = 2 * np.pi * np.linspace(0, 1, 5)
y = np.c_[np.cos(theta), np.sin(theta)]
cs = CubicSpline(theta, y, bc_type='periodic')
print("ds/dx={:.1f} ds/dy={:.1f}".format(cs(0, 1)[0], cs(0, 1)[1]))
x=0.0 ds/dy=1.0
xs = 2 * np.pi * np.linspace(0, 1, 100)
plt.figure(figsize=(6.5, 4))
plt.plot(y[:, 0], y[:, 1], 'o', label='data')
plt.plot(np.cos(xs), np.sin(xs), label='true')
plt.plot(cs(xs)[:, 0], cs(xs)[:, 1], label='spline')
plt.axes().set_aspect('equal')
plt.legend(loc='center')
plt.show()
```

In the following screenshot, we can see the final result:

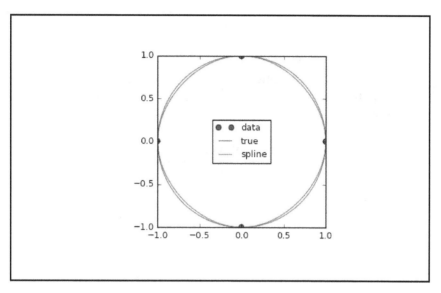

Defining a B-spline for a given set of control points

In the next section, we will look at how to solve B-splines given some controlled data.

Getting ready

We need to follow some instructions and install the prerequisites.

How to do it...

1. Univariate the spline in the B-spline basis
2. Execute the following: $S(x)=\sum j=0n-1cjBj,k;t(x)S(x)=\sum j=0n-1cjBj,k;t(x)$
3. Where it's $Bj,k;tBj,k;t$ are B-spline basis functions of degree k and knots t

4. We can use the following parameters:

Parameters	`t:` `ndarray, shape (n+k+1,)` knots. `c:` `ndarray, shape (>=n, ...)`. Spline coefficients. `k:` `int`. B-spline order. `extrapolate:` `bool`, optional. Whether to extrapolate beyond the base interval, t[k] .. t[n], or to return `NaN` values. If `True`, extrapolates the first and last polynomial pieces of B-spline functions active on the base interval. Default is `True`. `axis:` `int`; optional. Interpolation axis. The default is zero.

How it works ...

Here, we construct a quadratic spline function on the base interval `2 <= x <= 4` and compare it with the naive way of evaluating the spline:

```
from scipy import interpolate
import numpy as np
import matplotlib.pyplot as plt

# sampling
x = np.linspace(0, 10, 10)
y = np.sin(x)

# spline trough all the sampled points
tck = interpolate.splrep(x, y)
x2 = np.linspace(0, 10, 200)
y2 = interpolate.splev(x2, tck)

# spline with all the middle points as knots (not working yet)
# knots = x[1:-1] # it should be something like this
knots = np.array([x[1]]) # not working with above line and just seeing what
this line does
weights = np.concatenate(([1],np.ones(x.shape[0]-2)*.01,[1]))
tck = interpolate.splrep(x, y, t=knots, w=weights)
x3 = np.linspace(0, 10, 200)
y3 = interpolate.splev(x2, tck)

# plot
plt.plot(x, y, 'go', x2, y2, 'b', x3, y3,'r')
plt.show()
```

Note that outside of the base interval, results differ. This is because `BSpline` extrapolates the first and last polynomial pieces of B-spline functions active on the base interval.

This is the result of solving the problem:

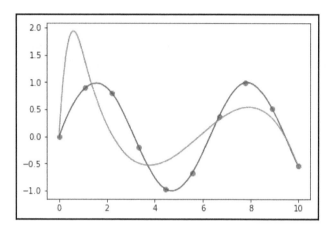

Differentiation

In the following section, we will look at how to solve differentiation and some of its applications.

Getting ready

There are three ways to approach the computation of derivatives:

- Numerical differentiation refers to the process of approximation of the derivative of a given function at a point. In SciPy, we have the following procedures, which will be covered in detail:

    - For generic univariate functions, there is the central difference formula with fixed spacing.

    - It is always possible to perform numerical differentiation via Cauchy's theorem, which transforms the derivative into a definite integral. This integral is then treated with the techniques of numerical integration explained before.

- Symbolic differentiation refers to computation of functional expressions of derivatives of functions, pretty much in the same way as we would do manually. It is termed symbolic because unlike its numerical counterpart, symbols take the roles of variables rather than numbers or vectors of numbers. To perform symbolic differentiation, we require a **computer algebra system** (**CAS**), and in the SciPy stack, this is achieved mainly through the SymPy library (see `http://docs.sympy.org/latest/index.html`). Symbolic differentiation and posterior evaluation are good substitutes for numerical differentiation for very basic functions. However, in general, this method leads to overcomplicated and inefficient code. The speed of purely numerical differentiation is preferred, in spite of the possible occurrence of errors.

- Automatic differentiation is another set of techniques to numerically evaluate the derivative of a function. It is not based on any approximation schema. This is without a doubt the most powerful option in the context of high derivatives of multivariate functions.

In the SciPy stack, this is performed through different unrelated libraries. Some of the most reliable are Theano (`http://deeplearning.net/software/theano/`) and FuncDesigner (`http://www.openopt.org/FuncDesigner`). For a comprehensive description and analysis of these techniques, a very good resource can be found at: `http://alexey.radul.name/ideas/2013/introduction-to-automatic-differentiation/`.

How to do it...

These are some types of differentiation that we will follow in this recipe:

- Numerical differentiation.
- The most basic scheme for numerical differentiation is performed with the central difference formula with uniformly spaced nodes. To maintain symmetry, an odd number of nodes is required to guarantee smaller round-off errors. An implementation of this simple algorithm is available in the `scipy.misc` module.

For information about the module `scipy.misc` and enumeration of its basic routines, refer to the online documentation at: `http://docs.scipy.org/doc/scipy-0.13.0/reference/misc.html`.

- To approximate the first and second derivatives of the polynomial function, for example, $f(x) = x^5$ at $x=1$ with 15 equally spaced nodes (centered at $x=1$) at distance $dx=1e-6$, we could issue the following command:

```
In [1]: import numpy as np
In [2]: from scipy.misc import derivative
In [3]: def f(x): return x**5
In [4]: derivative(f, 1.0, dx=1e-6, order=15)
Out[4]: 4.9999999997262723
In [5]: derivative(f, 1.0, dx=1e-6, order=15, n=2)
Out[5]: 19.998683310705456
```

Somewhat accurate, yet disappointing since the actual values are 5 and 20, respectively.

 Another flaw with this method (at least with respect to the implementation coded in SciPy) is the fact that the result relies on possibly large sums, and these are not stable. As users, we could improve matters by modifying the loop in the source of `scipy.misc.derivative` with the Shewchuk algorithm, for instance.

Symbolic differentiation

Exact differentiation for polynomials can be achieved using the `numpy.polynomial` module.

How it works...

```
In [6]: p = np.poly1d([1,0,0,0,0,0]); \
   ...: print (p)
   5
1 x
In [7]: np.polyder(p,1)(1.0)
In [7]: p.deriv()(1.0)
Out[7]: 5.0 Out[7]: 5.0
In [8]: np.polyder(p,2)(1.0)
In [8]: p.deriv(2)(1.0)
Out[8]: 20.0
Out[8]: 20.0
```

Symbolic differentiation is another way to achieve exact results:

```
In [9]: from sympy import diff, symbols
In [10]: x = symbols('x', real=True)
In [11]: diff(x**5, x)
In [12]: diff(x**5, x, x)
Out[11]: 5*x**4
Out[12]: 20*x**3
In [13]: diff(x**5, x).subs(x, 1.0)
Out[13]: 5.00000000000000
In [14]: diff(x**5, x, x).subs(x, 1.0)
Out[14]: 20.0000000000000
```

Note the slight improvement (both in notation and simplicity of coding) when we differentiate more involved functions than simple polynomials. For example, for $g(x) = e^{-x} sinx$ at $x=1$:

```
In [15]: def g(x): return np.exp(-x) * np.sin(x)
In [16]: derivative(g, 1.0, dx=1e-6, order=101)
Out[16]: -0.11079376536871781
In [17]: from sympy import sin as Sin, exp as Exp
In [18]: diff(Exp(-x) * Sin(x), x).subs(x, 1.0)
Out[18]: -0.110793765306699
```

A great advantage of symbolic differentiation over its numerical or automatic counterparts is the possibility to compute partial derivatives with extreme ease. Let's illustrate this point by calculating a fourth partial derivative of the multivariate function $h(x,y,z) = e^{xyz}$ at $x=1$, $y=1$, and $z=2$:

```
In [19]: y, z = symbols('y z', real=True)
In [20]: diff(Exp(x * y * z), z, z, y, x).subs({x:1.0, y:1.0, z:2.0})
Out[20]: 133.003009780752
```

Solving a one-dimensional ordinary differential equation

In the next section, we will look at how to solve one-dimensional ordinary differential equations.

Getting ready

We will need to follow some instructions and install the prerequisites.

How to do it...

1. As with integration, SciPy has some extremely accurate general-purpose solvers for systems of ordinary differential equations of the first order:

$$\frac{dy}{dt} = f(t, y), \ \ y(t) = (y_1(t), \ldots, y_n(t)), t \in \mathbb{R}$$

2. For real-valued functions, we have basically two flavors: ode (with options passed with the `set_integrator` method) and odeint (simpler interface). The syntax of ode is as follows:

   ```
   ode(f,jac=None)
   ```

3. The first parameter, `f`, is the function to be integrated, and the second parameter, `jac`, refers to the matrix of partial derivatives with respect to the dependent variables (the Jacobian). This creates an ode object with different methods to indicate the algorithm to solve the system (`set_integrator`), the initial conditions (`set_initial_value`), and different parameters to be sent to the function or its Jacobian.

4. The options for the integration algorithm are vode for the real-valued variable coefficient ODE solver, with fixed-leading-coefficient implementation (it provides Adam's method for non-stiff problems and BDF for stiff); zvode for the complex-valued variable coefficient ODE solver, with similar options to the preceding option; dopri5 for a **Runge-Kutta** method of order (4)5; dop853 for a Runge-Kutta method of order 8(5, 3).

5. The next code snippet presents an example of using `scipy.integrate.ode` to solve the initial value problem using the following formula:

$$y' = -20y, \ \ y(0) = 1$$

   ```
   from scipy.integrate import ode

   y0, t0 = [1.0j, 2.0], 0
   ```

```
def f(t, y, arg1):
    return [1j*arg1*y[0] + y[1], -arg1*y[1]**2]
def jac(t, y, arg1):
    return [[1j*arg1, 1], [0, -arg1*2*y[1]]]

r = ode(f, jac).set_integrator('zvode', method='bdf',
with_jacobian=True)
r.set_initial_value(y0, t0).set_f_params(2.0).set_jac_params(2.0)
t1 = 10
dt = 1
while r.successful() and r.t < t1:
    r.integrate(r.t+dt)
    print("%g %g" % (r.t, r.y))
```

How it works...

We compute each step sequentially and compare it to the actual solution, which is known. You will notice that there is virtually no difference:

```
import numpy
from scipy.integrate import ode
f=lambda t,y: -20*y # The ODE
actual_solution=lambda t:numpy.exp(-20*t) # actual solution
dt=0.01 # time step
solver=ode(f).set_integrator('dop853') # solver
solver.set_initial_value(1,0) # initial value
while solver.successful() and solver.t<=1+dt:
    # solve the equation at successive time steps,
    # until the time is greater than 1
    # but make sure that the solution is successful
        print (solver.t, solver.y, actual_solution(solver.t))
    # We compare each numerical solution with the actual
    # solution of the ODE
        solver.integrate(solver.t + dt) # solve next step
```

Solving a system of ordinary differential equations

In the following section, we will look at how to solve a system of ordinary differential equations.

Getting ready

We will need to follow some instructions and install the prerequisites.

How to do it...

These are the arguments for having this parameter on the functions:

1. To execute the following parameters, we just need to know the parameters of the equations.
2. Execute them with the following parameters:

Parameters	fun: Callable. Right-hand side of the system. The calling signature is fun(x, y), or fun(x, y, p) if parameters are present. All arguments are ndarray x with shape (m,), and y with shape (n, m), meaning that y[:, i] corresponds to x[i] and p with shape (k,). The return value must be an array with shape (n, m) and with the same layout as y. bc: Callable. A function evaluating residuals of the boundary conditions. The calling signature is bc(ya, yb) or bc(ya, yb, p) if parameters are present. All arguments are ndarray ya and yb with shape (n,) and p with shape (k,). The return value must be an array with shape (n + k,). x: array_like, shape (m,). Initial mesh. Must be a strictly increasing sequence of real numbers with x[0]=a and x[-1]=b. y: array_like, shape (n, m). Initial guess for the function values at the mesh nodes, i^{th} column corresponds to x[i]. For problems in a complex domain, pass *y* with a complex data type (even if the initial guess is purely real). p: array_like with shape (k,) or none, optional. Initial guess for the unknown parameters. If it is none (default), it is assumed that the problem doesn't depend on any parameters. S: array_like with shape (n, n) or none. A matrix defining the singular term. If is is none (default), the problem is solved without the singular term.

	fun_jac: Callable or none, optional. A function computing the derivatives of f with respect to y and p. The calling signature is fun_jac(x, y), or fun_jac(x, y, p) if parameters are present. The return must contain 1 or 2 elements in the following order: • df_dy: array_like with shape (n, n, m) where an element (i, j, q) equals d f_i(x_q, y_q, p) / d (y_q)_j • df_dp: array_like with shape (n, k, m) where an element (i, j, q) equals d f_i(x_q, y_q, p) / d p_j
	Here, q numbers the nodes at which x and y are defined, whereas i and j number the vector components. If the problem is solved without unknown parameters, df_dp should not be returned. If fun_jac is none (default), the derivatives will be estimated by the forward finite differences. bc_jac: Callable or none, optional. A function computing the derivatives of bc with respect to ya, yb and p. The calling signature is bc_jac(ya, yb), or bc_jac(ya, yb, p) if parameters are present. The return must contain 2 or 3 elements in the following order: • dbc_dya: array_like with shape (n, n) where an element (i, j) equals d bc_i(ya, yb, p) / d ya_j • dbc_dyb: array_like with shape (n, n) where an element (i, j) equals d bc_i(ya, yb, p) / d yb_j • dbc_dp: array_like with shape (n, k) where an element (i, j) equals d bc_i(ya, yb, p) / d p_j If the problem is solved without unknown parameters, dbc_dp should not be returned. If bc_jac is none (default), the derivatives will be estimated by the forward finite differences. tol: float, optional. The desired tolerance of the solution. If we define r = y' − f(x, y) where y is the found solution, then the solver tries to achieve on each mesh interval norm(r / (1 + abs(f)) < tol, where norm is estimated in a root mean squared sense (using a numerical quadrature formula). The default is 1e-3. max_nodes: int, optional. The maximum allowed number of the mesh nodes. If exceeded, the algorithm terminates. The default is 1000. verbose: {0, 1, 2}, optional. The level of algorithm's verbosity: • 0 (default): Work silently • 1: Display a termination report • 2: Display progress during iterations

Returns	A bunch object with the following fields defined: `sol`: `PPoly`. Found solution for `y` as `scipy.interpolate.PPoly` instance, a C1 continuous cubic spline. `p`: `ndarray` or None, shape `(k,)`. Found parameters. None, if the parameters were not present in the problem. `x`: `ndarray`, shape `(m,)`. The nodes of the final mesh. `y`: `ndarray`, shape `(n, m)`. Solution values at the mesh nodes. `yp`: `ndarray`, shape `(n, m)`. Solution derivatives at the mesh nodes. `rms_residuals`: `ndarray`, shape `(m - 1,)`. RMS values of the relative residuals over each mesh interval (see the description of the `tol` parameter). `niter`: `int`. The number of completed iterations. `status`: `int`. The reason for algorithm termination: • `0`: The algorithm converged to the desired accuracy • `1`: The maximum number of mesh nodes was exceeded • `2`: A singular Jacobian encountered when solving the collocation system `message`: `String`. Verbal description of the termination reason. `success`: `bool`. `True` if the algorithm is converged to the desired accuracy (`status=0`).

How it works...

Here, we have an one example to give away the details:

```
from scipy.integrate import solve_bvp
res_a = solve_bvp(fun, bc, x, y_a)
res_b = solve_bvp(fun, bc, x, y_b)
x_plot = np.linspace(0, 1, 100)
y_plot_a = res_a.sol(x_plot)[0]
y_plot_b = res_b.sol(x_plot)[0]

import matplotlib.pyplot as plt
plt.plot(x_plot, y_plot_a, label='y_a')
plt.plot(x_plot, y_plot_b, label='y_b')
plt.legend()
plt.xlabel("x")
plt.ylabel("y")
plt.show()
```

This is the result of the execution:

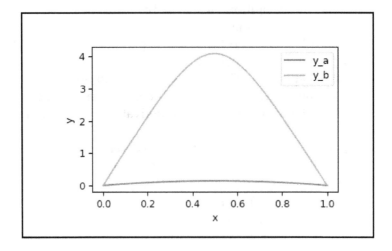

This is the code of the other solution:

```
x_plot = np.linspace(0, 1, 100)
y_plot = res_b.sol(x_plot)[1]
plt.plot(x_plot, y_plot)
plt.xlabel("x")
plt.ylabel("y")
plt.show()
```

The final result is shown in the next screenshot:

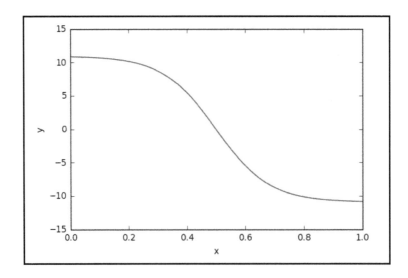

Solving differential equations and systems with parameters

In the following section, we will see how to solve differential equations and systems with parameters.

Getting ready

We will need to have the package and install the prerequisites `spicy.integrate`.

How to do it...

No book on scientific computing is complete without revisiting Lorenz attractors. SciPy excels both at the computation of solutions and presentation of ideas based on systems of differential equations, and we will show how and why in this recipe:

- Consider a two-dimensional fluid cell that is heated from underneath and cooled from above, much like what occurs with the Earth's atmosphere. This creates convection which can be modeled by a single partial differential equation, for which a decent approximation has the form of the following system of ordinary differential equations:

$$\begin{cases} \frac{dx}{dt} = \sigma(y - x) \\ \frac{dy}{dt} = rx - y - xz \\ \frac{dz}{dt} = xy - bz \end{cases}$$

1. The variable x represents the rate of convective overturning. The variables y and z stand for the horizontal and vertical temperature variations, respectively. This system depends on four physical parameters, the descriptions of which are far beyond the scope of this book. The important point is that we may model the Earth's atmosphere with these equations, and in that, case a good choice for the parameters is given by *sigma = 10.0*, and *b = 8/3.0*. For certain values of the third parameter, we have systems for which the solutions behave chaotically. Let's explore this effect with the help of SciPy.

2. In the following code snippet, we will use one of the solvers in the `scipy.integrate`module, as well as the plotting utilities:

```
from scipy.integrate import ode

y0, t0 = [1.0j, 2.0], 0

def f(t, y, arg1):
    return [1j*arg1*y[0] + y[1], -arg1*y[1]**2]
def jac(t, y, arg1):
    return [[1j*arg1, 1], [0, -arg1*2*y[1]]]
```

How it works...

This is the main code of the solution:

```
import numpy
from numpy import linspace
import scipy
from scipy.integrate import odeint
import matplotlib.pyplot as plt
from mpl_toolkits.mplot3d import Axes3D
sigma=10.0
b=8/3.0
r=28.0
f = lambda x,t: [sigma*(x[1]-x[0]), r*x[0]-x[1]-x[0]*x[2], x[0]*x[1]-
b*x[2]]
```

This is the final result:

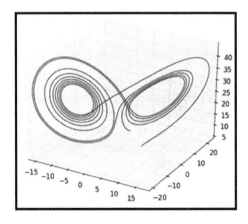

This precisely shows the chaotic behavior of the solutions. Let's observe the fluctuations of the vertical temperature in detail, along with the fluctuation of the horizontal temperature against the vertical. Issue the following commands:

```
plt.rcParams['figure.figsize'] = (10.0, 5.0)
plt.subplot(121); plt.plot(t,Z)
plt.subplot(122); plt.plot(Y,Z)
plt.show()
```

This produces the following plots that show vertical temperature with respect to time (left-hand side plot) and horizontal versus vertical temperature (right-hand side plot):

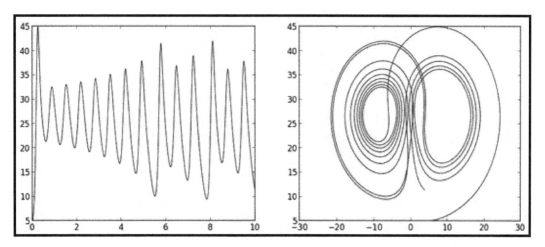

Using ode and the objected-oriented interface to solve differential equations

In the following section, we will see how to use ode and the object-oriented interface and how to solve differential equations using it.

Getting ready

We will need to have the prerequisites, such as `scipy.integrate.odeint`.

How to do it...

Some parameters are included in the following tables; we need to execute them:

1. Get the function accordingly
2. Name the following parameters
3. Execute it
4. Get the solution

Parameters	func: `callable(y, t0, ...)`. This computes the derivative of y at t0. y0: Array. The initial condition on y (can be a vector). t: Array. A sequence of time points for which to solve for y. The initial value point should be the first element of this sequence. args: Tuple, optional. Extra arguments to pass to the function. Dfun: `callable(y, t0, ...)`. The gradient (Jacobian) of func. col_deriv: `bool`, optional. `True` if Dfun defines derivatives down columns (faster); otherwise Dfun should define derivatives across rows. full_output: `bool`, optional. `True` if to return a dictionary of optional outputs as the second output printmessg: `bool`, optional. It denotes whether to print the convergence message or not.
Returns	y: Array, shape (`len(t), len(y0)`). An array containing the value of y for each desired time in t, with the initial value y0 in the first row. infodict: `dict`; this is returned only if full_output == True.

We have some additional information in terms of the output, which is mostly self explanatory. You will love to know about it. Please refer to the following:

Key	Meaning
hu	Vector of step sizes successfully used for each time step.
tcur	Vector with the value of t reached for each time step. Will always be at least as large as the input times.
tolsf	Vector of tolerance scale factors, greater than 1.0, computed when a request for too much accuracy was detected.
tsw	Value of t at the time of the last method switch (given for each time step).
nst	Cumulative number of time steps.

nfe	Cumulative number of function evaluations for each time step.
nje	Cumulative number of Jacobian evaluations for each time step.
nqu	A vector of method orders for each successful step.
imxer	Index of the component of largest magnitude in the weighted local error vector (e / ewt) on an error return; it is −1 otherwise.
lenrw	The length of the double work array required.
leniw	The length of integer work array required.
mused	A vector of method indicators for each successful time step: 1: adams (nonstiff), 2: bdf (stiff).

How it works ...

The solution is an array with shape (101, 2). The first column is theta(t), and the second is omega(t). The following code plots both components.

We can see it in the following code:

```
import matplotlib.pyplot as plt
plt.plot(t, sol[:, 0], 'b', label='theta(t)')
plt.plot(t, sol[:, 1], 'g', label='omega(t)')
plt.legend(loc='best')
plt.xlabel('t')
plt.grid()
plt.show()
```

The final result is this:

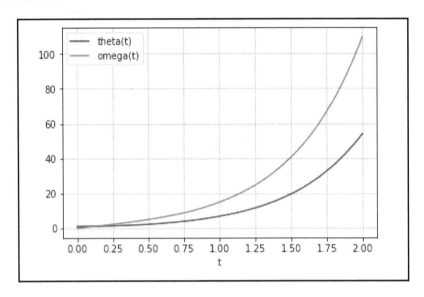

9
Statistics and Probability

In this chapter, we will present recipes to perform the following tasks:

- Computing the probability mass function of a discrete random variable
- Computing the probability density function of a continuous random variable
- Computing the cumulative distribution function for a random variable
- Computing the values of inverse probabilities associated with a random variable
- Computing the average, standard deviation, and higher moments of a distribution
- Computing probabilities associated with the multivariate Gaussian distribution
- Computing the summary statistics of a dataset

Introduction

In the previous chapters, we have become familiar with instantiating arrays, working on data frames and matrices, solving equations, creating special functions, and performing calculus using functions in SciPy. Together with all this, in a wide variety of tasks, we have assessed the probability of certain events happening or performing statistical analysis on top of data.

Statistics is a branch of mathematics dealing with the collection, analysis, interpretation, and presentation of masses of numerical data.

Furthermore, probability is the extent to which an event is likely to occur, which is measured using the ratio of the favorable cases to the whole number of cases possible.

This chapter shows you how to work with the statistical and probability tools available in SciPy.

There are multiple applications in which probability and statistics play a vital role. Some of those applications are as follows:

- Predicting the likelihood of an event happening
- Forecasting the value of a certain variable over time
- Image compression algorithms
- Business analytics, which deals with collecting, analyzing, and deriving insights from data to support the formulation of strategies
- Statistical quality control to identify outliers from a given output/outcome
- Signal processing
- Operational research

While these applications are some of the major applications for probability and statistics, there are multiple other applications where probability and statistics form the backbone of the analysis.

Computing the probability mass function of a discrete random variable

A random variable is a variable whose value is unknown, or a variable for which the value changes over different iterations of the experiment.

For example, when we roll a die, the outcome of rolling the dice will vary over different iterations and hence the outcome becomes a random variable.

A random variable is discrete if the outcome of the random variable is limited to a few possible outcomes.

For example, the outcome of rolling a fair dice can only be 1, 2, 3, 4, 5, or 6; it cannot be a number beyond that. Thus, the outcome is limited to only a few possible values.

In the previous example, whenever a die is rolled, there is a probability associated with each outcome. For example, if a fair dice is rolled once, the probability that the outcome is 4 is 1/6, as all outcomes have an equal chance of being obtained.

 A **probability mass function (pmf)** is a function that provides the probability that a discrete random variable is exactly equal to some value.

Given that the pmf assigns a probability number to each possible outcome of the experiment, we can think of it as the pmf assigning a mass (weight) to each possible outcome, where the mass is high if the likelihood of occurrence (probability) is high.

There can be multiple scenarios to specify a pmf:

- **Binomial discrete distribution**: Calculating the probability of the outcome when there are only two possible outcomes
- **Multivariate discrete distribution**: Calculating the probability when there are multiple possible outcomes

Binomial discrete distribution

In order to understand binomial discrete distribution, let us consider the following example:

What is the probability that, when a fair coin is tossed twice, it lands on heads only once?

For the preceding scenario, we would look at all the possible outcomes of the two experiments (where tossing a coin once is one experiment).

We have listed all the possible outcomes of the two coin tosses (two experiments), as follows:

$$\{HH, HT, TH, TT\}$$

Given that, there are four possible outcomes of the two experiments, of which only two outcomes satisfy the criterion that we laid out (the coin lands on heads only once out of the two experiments). The probability of the event happening is *2/4 = 1/2 = 0.5.*

Mathematically, this can be represented in the following way. The pmf of a binomial random variable X is:

$$f(x) = \binom{n}{x} p^x (1-p)^{n-x}$$

The following applies:

- *f(x)* is the probability of an event happening
- *n* is the number of experiments
- *x* is the number of times a condition has to be satisfied
- *p* is the probability of an event happening

If we replicate the preceding laid-out example using the formula, we obtain the following:

- *n=2*
- *x=1*
- *p=1/2*

This translates to *f(x)* being equal to 1/2.

Multinomial discrete distribution

In order to understand multinomial discrete distribution, let's consider the following scenario:

In a football match, the outcome of the game is equally likely to be a win, draw, or loss for a given team. What is the probability that if the two teams play twice, there will be exactly one win and one draw for the team?

Similar to the way in which we solved the previous example, let us consider all the possible outcomes of playing two games (conducting two experiments):

$$\{WW, WD, WL, DD, DW, DL, LL, LD, LW\}$$

From all the possible outcomes listed, one win and one draw can happen in only two possible outcomes (*DW* and *WD*).

Thus, the probability of the event happening is 2/9.

Mathematically, the preceding scenario is solved as follows:

$$P(X_1 = x_1, \ldots, X_n = x_n) = \frac{N!}{\prod_{i=1}^{n} x_i!} \prod_{i=1}^{n} \theta_i^{x_i}$$

The following applies:

- N is the number of times an experiment is conducted
- x_i is the number of times each event is expected to happen
- P is the probability associated with an event happening

The formula translates into the following computation:

- *P(X1 = 1W) = 1/3*, as all the three outcomes are equally likely
- *P(X2 = 1D) = 1/3*
- *N = 2*, as the experiment is conducted twice
- *x1 = 1* and *x2 = 1*, as both win and draw are expected to occur twice

All these numbers translate into the following:

$$2!/(1! * 1!) * ((1/3)^1 * (1/3)^1 * (1/3)^0) = 2/9$$

The preceding functions will be implemented in Python in the following section.

How to do it...

The pmf of a binomial outcome is obtained in Python using a set of functions in the `binom` module within `scipy.stats`.

The steps to plot the probability mass function of a discrete random variable are as follows:

1. Import the relevant packages:

```
from scipy.stats import binom
import matplotlib.pyplot as plt
import numpy as np
```

2. Specify the total number of experiments and the probability of an event occurring:

```
n=5
p=0.4
```

3. Extract the probability of an event occurring in *x* occasions out of the *n* experiments.

   Define *x*. In our case, *x* can take values from 0 to the maximum (which is *n*):

```
x=np.arange(0,(n+1))
print(x)
[0 1 2 3 4 5]
```

   Pass x through the pmf function. Note that the pmf function calculates the *f(x)* value as discussed in the previous section:

```
binom.pmf(x, n, p)
array([ 0.07776,  0.2592 ,  0.3456 ,  0.2304 ,  0.0768 ,  0.01024])
```

The probability of each value of x happening for the given n and p is given in the preceding output.

Now the output is obtained, let's do a few cross-checks to see if the output obtained is as per our expectations.

1. The sum of the probabilities of all the values of x combined should be equal to 1:

```
np.sum(binom.pmf(x, n, p))
0.99999999999999978
```

2. Check the probability of x=1 using the formula listed previously:

$$5C1 * (0.4)^1 * (0.6)^4 = 0.2592$$

The output is equal to the output obtained when passing x through the pmf function.

Visualizing the probability mass function

pmf gives us an output of the mass (probability) associated with each likely output. In the example that we looked at earlier, we have the probability associated with each value of x (although the probabilities vary depending on the value of x).

In this section, we will look at plotting the probability mass function for each possible value of x:

1. Import the relevant packages for plotting:

```
import matplotlib.pyplot as plt
```

2. Initialize the plots:

```
fig, ax = plt.subplots(1, 1)
```

3. Plot the values of x along with the probability mass function associated with each value of x:

```
ax.plot(x, binom.pmf(x, n, p), 'bo', ms=8, label='binom pmf')
[<matplotlib.lines.Line2D at 0x681f474518>]
```

4. Show the plot:

```
plt.show()
```

The preceding code results in the following output:

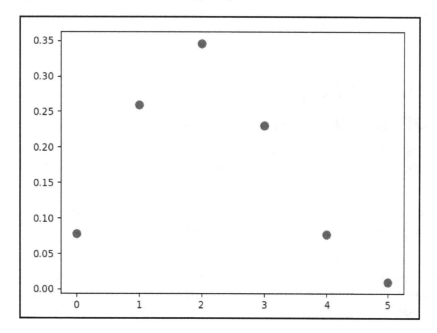

So far, we have looked at a random variable that has a binomial output. The code snippets in the following steps helps us to calculate the pmf of a variable that has a multinomial output:

1. Import the relevant packages:

```
from scipy.stats import multinomial
```

2. Initialize the values of N and the probability associated with each outcome:

```
rv = multinomial(8, [0.3, 0.2, 0.5])
```

Note that, in the preceding code, N=8 and the probabilities associated with each outcome are [0.3, 0.2, 0.5].

3. Calculate the probability mass function for the three outcomes to appear using a given set of values:

```
rv.pmf([1, 3, 4])
0.042000000000000072
```

Note that the values [1, 3 ,4] represent the number of times the first outcome is expected to occur (once), the number of times the second outcome is expected to occur (thrice), and the number of times the third outcome is expected to occur (four times).

Computing the probability density function of a continuous random variable

In the previous section, we saw how to calculate the pmf of a discrete random variable. In this section, we will calculate the **probability density function (pdf)** of a continuous random variable.

In order to understand pdf better, we will look at a toy example. Let us take a scenario where we are considering John—a student—and his time of arrival for a class.

In the previous section, we looked at a discrete scenario—*John could be early to class or late to class*. In this section, we will be considering the magnitude of how early to class or late to class John may arrive in minutes. So, we will translate the problem set from a discrete outcome (late or early) to a continuous outcome (magnitude of minutes that he was early to class).

Computationally, to go from discrete to continuous we simply replace sums with integrals.

This can be visually represented in charts, as follows:

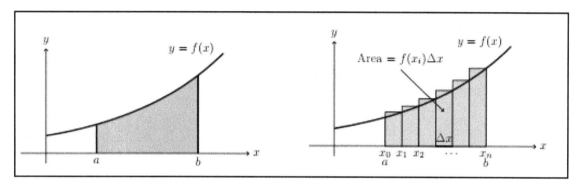

Let's say we would like to calculate the area of the shaded region in the left chart from **a** to the **b** on the x axis.

The area calculation can be achieved by breaking down the larger shaded region on the left chart into rectangles that have very small widths on the x axis. Thus, the area of the shaded region now becomes:

$$area \approx sum\ of\ rectangle\ areas = f(x_1)\Delta x + f(x_2)\Delta x + \ldots + f(x_n)\Delta x = f(x_i)\Delta x$$

In the preceding scenario, the function *f(x)* is called a probability density function if it satisfies the following conditions:

- *f(x)>0*—that is, *f* is non-negative
- *∫f(x) dx = 1* (this is equivalent to *P(−∞ < X < ∞) = 1*)

How to do it...

In order to understand the process of generating a probability density function and calculating the associated probabilities, let us consider the following example:

A curve follows a normal distribution with mean 0 and standard deviation 1. What is the probability that the value of the normal distribution is between 0.5 and 1?

In order to solve this scenario, we need to perform the following steps:

1. Generate the numbers associated with the given pdf and see where most of the numbers fall in the distribution.
2. Assess the probability of the numbers being between 0.5 and 1 if they are generated as per the distribution.

A normal distribution with a mean of 0 and a standard deviation of 1 can be represented by the following formula:

$$\frac{np.\,exp(-(x)**2/2)}{np.\,sqrt(2*math.\,pi)}$$

Where *x* is the continuous random variable.

In the preceding formula, we can see that the output is never less than 0.

Let us go ahead and generate the numbers associated with this distribution:

1. Import the relevant packages:

```
import numpy as np
import math
from scipy.stats import norm
import scipy.stats as st
```

2. Generate the numbers and store them in a list:

```
r = norm.rvs(loc=0, scale=1,size=10000)
```

In the preceding line of code, we are generating 10,000 numbers using a normal distribution with a mean of 0 and a standard deviation of 1.

In the following snippet of code, let us try to understand how we are able to generate the 10,000 numbers that follow the standard normal distribution:

```
class my_pdf(st.rv_continuous):
    def _pdf(self,x):
        return np.exp(-((x)**2)/2)/np.sqrt(2*math.pi)
my_cv = my_pdf(a=-100, b=100, name='my_pdf')
```

In the preceding snippet, we are instantiating a class named `my_pdf`, which is a subclass of the `rv_continuous` class in `scipy.stats`.

The _pdf method overwrites the default `pdf` function to return the `pdf` function of interest, which is *np.exp(-((x)**2)/2)/np.sqrt(2*math.pi)* in this particular scenario.

In the final part of the code, we are specifying the range of values that the x values can take, which is between `-100` and `100`.

Numbers that follow the given distribution can now be generated using the `rvs` function.

3. Plot the numbers in terms of the frequency with which they occur in the given distribution:

```
plt.hist(r)
plt.show()
```

The resulting plot looks like the following:

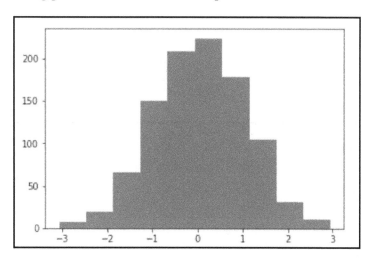

In the final step, let's go ahead and calculate the number of times the output is between `0.5` and `1` out of the total number of observations.

```
sum((r>0.5) & (r<1)) / len(r)
```

The output in the instance we ran in the code snippet is `15%`.

In order to validate the results, let us check the number of times the output is beyond 0—intuitively, this should be close to 50%, as our distribution has a mean of 0 and a standard deviation of 1:

```
sum((r>0))/len(r)
```

The output of the preceding calculation is 50.1%, validating our hypothesis.

An alternative way to plot the probability distribution is as follows:

1. Generate a normal distribution with a given mean and standard deviation:

```
gaussian = stats.norm(loc=0, scale=1.0)
```

2. Initialize the values of the continuous random variable x.

```
x = np.linspace(-5, 5, 1000)
```

3. Generate the pdf:

```
y = gaussian.pdf(x)
```

4. Plot the distribution:

```
plt.plot(x,y)
```

The result looks like the following:

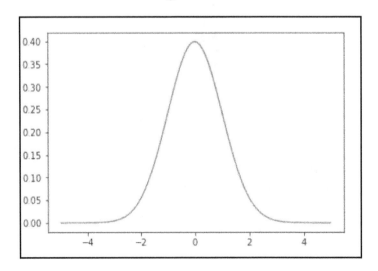

Computing the cumulative distribution function for a random variable

In the previous section, we looked at the probability distribution of a continuous random variable.

In the next section, we will consider the cumulative distribution of a continuous random variable.

How to do it...

The cumulative distribution function (CDF) of a random variable is calculated as follows:

The CDF of a continuous random variable is calculated in a way similar to that in which we calculate the pdf of a continuous random variable.

The following code snippet calculates the CDF of a variable:

1. Import the relevant packages:

```
import numpy as np
import matplotlib.pyplot as plt
from scipy.stats import norm
```

2. Calculate the CDF of the variable:

```
gaussian = norm(loc=0, scale=1.0)
x = np.linspace(-5, 5, 1000)
y = gaussian.cdf(x)
```

In the preceding snippet of code, we have calculated the CDF for the continuous random variable that is expected to generate a Gaussian distribution with a mean of 0 and a standard deviation of 1.

Note that we have generated 1,000 numbers that are uniformly distributed between −5 and 5.

Once the numbers are generated, we pass the numbers through the cdf function to determine the cumulative density.

3. Plot the CDF:

```
plt.plot(x,y)
```

In the following plot, we can see that most of the numbers are located close to 0 and hence there is a spike in the curve at around 0.

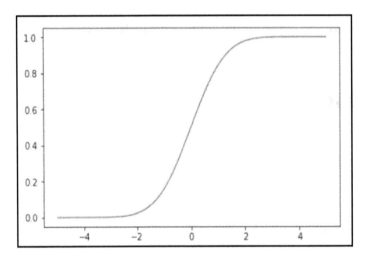

4. We can generate the probability that a number in the distribution will be less than a certain a value as follows:

```
gaussian.cdf(0)
```

The preceding code snippet gives us the cumulative probability of a number that follows the distribution and is less than 0—the output of which is 0.5.

The calculations can also be done in the following way:

1. Import the relevant packages:

```
from scipy import stats
import math
```

2. Overwrite the function to obtain a normal distribution:

```
class my_pdf(stats.rv_continuous):
    def _pdf(self,x):
        return np.exp(-((x)**2)/2)/np.sqrt(2*math.pi)
```

3. Specify the maximum and minimum values of the distribution:

```
my_cv = my_pdf(a=-5, b=5, name='my_pdf')
```

4. Generate the numbers that follow the preceding distribution by using the `rvs` function:

```
x=my_cv.rvs(size=10000)
```

5. Plot the histogram of the values obtained:

```
plt.hist(x)
plt.show()
```

The plot looks like the following:

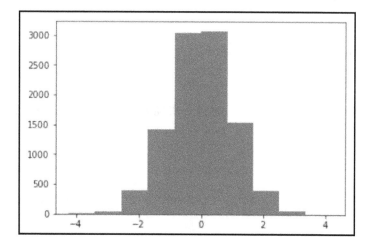

6. Calculate the number of observations that are less than 0 out of the total number of observations:

```
sum(x<0)/len(x)
```

The output of the preceding line of code is `0.50329999999999997`, which is very close to the expected 50% of observations that are less than `0`.

Computing the values of inverse probabilities associated with a random variable

The inverse probability of a random variable is the inverse of the CDF associated with the distribution.

The **percent point function** (**PPF**) gives us the value of the continuous random variable that is associated with the percent value (quantile value).

How to do it...

In order to understand this better, let us consider the following:

1. Import the relevant packages:

   ```
   from scipy.stats import norm
   ```

2. Extract the value associated with the 95% percentile (quantile) value:

   ```
   norm.ppf(0.95)
   ```

   The output of the preceding line of code is 1.6448536269514722.

3. The inverse of the preceding output can be calculated as follows:

   ```
   norm.cdf(1.6448)
   ```

   The output of this is 0.94999446890607997.

Computing the average, standard deviation, and higher moments of a distribution

In order to understand the way in which we can extract the average of a distribution, let us go through the following scenario.

How to do it...

We will initialize a normal distribution with a given average and standard deviation. Once initialized, we will consider the output that we should be expecting.

Initializing a normal distribution

A normal variable with a given mean and standard deviation can be initialized by using the rvs function in scipy.stats.norm:

1. Import the relevant packages:

   ```
   from scipy import stats
   ```

2. Initialize a variable with a given mean and standard distribution:

   ```
   x = stats.norm.rvs(loc=3, scale=2, size=(1000))
   ```

   In the preceding line of code, we have initialized a variable x, which has a mean value of 3, a standard deviation of 2, and a total size of 1,000.

   Note that the mean is referred to as loc and the standard deviation as scale in the line of code.

3. Now, we have obtained the values of x, let's go ahead and plot the distribution of x:

   ```
   import matplotlib.pyplot as plt
   %matplotlib inline
   plt.hist(x)
   plt.show()
   ```

In the preceding snippet of code, we have plotted the histogram of the values of x, and the plot looks like the following:

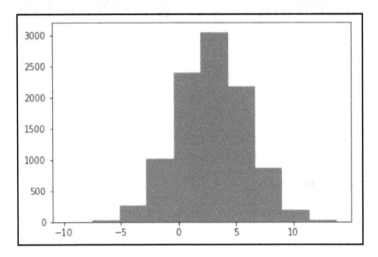

In the preceding plot, we can see that the distribution is centered around 3 (which is what we expect, as we initialized the distribution with a mean of 3).

Given that we have the distribution we expected, let us go ahead and calculate the average, standard deviation, and moments of the distribution.

Average and standard deviation of a distribution

The average of a distribution can be calculated by using the mean function within the numpy package. The way to extract the mean of the initialized distribution is as follows:

1. Import the relevant packages:

```
import numpy as np
```

2. Calculate the average of the earlier initialized distribution:

```
np.mean(x,axis=0)
```

In the preceding line of code, we have calculated the mean of all the values within x. The output of the preceding line of code is as follows:

```
3.0372443687293549
```

The output is close to what we expected, as we have initialized the distribution with a mean of three.

The output does not match three exactly, as we have taken a smaller sample size. Let us calculate the mean of a normal distribution in a similar way, but this time with a larger sample size:

```
from scipy import stats
x = stats.norm.rvs(loc=3, scale=3, size=(1000000))
np.mean(x,axis=0)
```

The output is 2.9963541802866853, which is a value that is very close to three—the expected mean value.

3. Calculate the standard deviation of the earlier initialized distribution using the std function:

```
np.std(x)
```

The output is 3.002600053097471, which is very close to the standard deviation of the distribution that we expected.

Calculating the moments of a distribution

A moment is a specific quantitative measure of the shape of a set of points. It is often used to calculate coefficients of skewness and kurtosis.

The kth moment of a distribution is calculated with the following formula:

$$m_k = \frac{1}{n} \sum_{i=1}^{n} (x_i - \overline{x})^k$$

Note that the preceding formula is equivalent to a mean-shifted average when *k=1*.

Similarly, the same preceding formula is equivalent to the variance of an array when *k=2*.

Furthermore, *k=3* represents the skewness and *k=4* represents the kurtosis of the array.

In the following section, let us look at calculating the kth moment of a distribution:

1. Import the relevant packages:

```
from scipy.stats import moment
```

2. Initialize a distribution.

   In order to understand how moment functions work, let us initialize a simple distribution so that we can easily interpret the results:

   ```
   x=np.array([1,2,3,4,5])
   ```

   In the preceding code snippet, we have initialized an array with numbers from 1 to 5.

3. Calculate the kth moment of the distribution.

   The kth moment of a distribution is calculated by using the moment function in scipy.stats:

   ```
   moment(x,moment=1)
   ```

   In the preceding code snippet, we have calculated the first moment of the distribution, which is equivalent to the average of the distribution.

   The output of the preceding code snippet is 0.0, which is the mean-shifted average of the original distribution.

4. Similarly, the output of the following code snippet is equivalent to the variance of the distribution:

   ```
   moment(x,moment=2)
   ```

   The output of the preceding line of code is equal to 2.0, which is the same as the output when the variance of the distribution is calculated:

   ```
   np.var(x)
   ```

5. The skewness of the distribution can be calculated as follows:

```
moment(x,moment=3)
```

   The output of the preceding snippet of code is 0.0. It is intuitive to us because the distribution is equally distributed on both sides of the mean, and hence there is relatively very little, if any, skewness.

6. The kurtosis of a distribution can be calculated as follows:

```
moment(x,moment=4)
```

   The output is 6.7999999999999998.

   Positive kurtosis indicates that very few data points are located in the tails.

Computing probabilities associated with the multivariate Gaussian distribution

A **multivariate Gaussian distribution** is a generalization of the one-dimensional (univariate) normal distribution to higher dimensions. A random vector is said to be *k*-variate normally distributed if every linear combination of its *k* components has a univariate normal distribution.

Before looking at the multivariate Gaussian distribution, let's consider the univariate distribution.

A univariate distribution is generated with the following formula:

$$g(x) = \frac{1}{\sigma\sqrt{2\pi}} e^{-\frac{1}{2}\left(\frac{x-\mu}{\sigma}\right)^2}$$

In the preceding formula, the following applies:

- σ represents the standard deviation of the distribution
- μ represents the mean of the distribution

Given the preceding two parameters, a Gaussian distribution with a certain mean and standard deviation is generated by varying the values of x from -∞ to ∞.

A typical plot of a Gaussian distribution for different values of mean and standard deviation can look as follows:

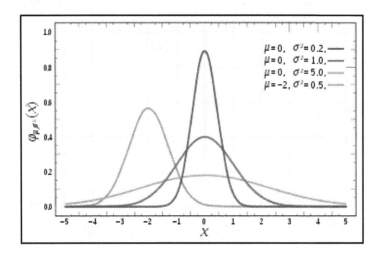

A multivariate gaussian distribution is very similar to a univariate Gaussian distribution. However, given that there are multiple variables involved, a univariate mean (the mean of a single variable) is replaced by a multivariate mean (the mean of each of the variables), and the variance of a single variable is replaced by the covariance of the multiple variables present in a multivariate Gaussian distribution.

Thus, the formula of a multivariate Gaussian distribution is as follows:

$$P(x_1, x_2) = \frac{1}{2\pi\sigma_1\sigma_2\sqrt{1-\rho^2}} exp\left[-\frac{z}{2(1-\rho^2)}\right],$$

$$where$$

$$z \equiv \frac{(x_1 - \mu_1)^2}{\sigma_1^2} - \frac{2\rho(x_1 - \mu_1)(x_2 - \mu_2)}{\sigma_1\sigma_2} + \frac{(x_2 - \mu_2)^2}{\sigma_2^2}$$

$$and$$

$$\rho \equiv cor(x_1, x_2) = \frac{v_{12}}{\sigma_1\sigma_2}$$

A typical chart of a bivariate normal distribution for a given value of x can look like the following:

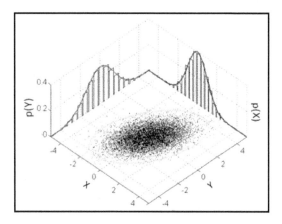

A k-variate multinormal distribution is a generalization of the bivariate normal distribution. The k-multivariate distribution with mean vector μ and covariance matrix $\sum$ is denoted by the following:

$$f_X(x, \ldots, x_k) = \frac{exp(-\frac{1}{2}(x-\mu))^T \sum^{-1}(x-\mu))}{\sqrt{(2\pi)^k |\sum|}}$$

$$= \frac{exp(-\frac{1}{2}(x-\mu))^T \sum^{-1}(x-\mu))}{\sqrt{|2\pi \sum|}}$$

In the following section, we will look at calculating the probability associated with a multivariate Gaussian distribution.

How to do it...

The probability of a given value in a multivariate Gaussian distribution is calculated as follows:

1. Import the relevant packages:

```
from scipy.stats import multivariate_normal
import numpy as np
```

2. Initialize the array of variables:

```
x = np.array([[1,2], [3,4]])
```

In the preceding code snippet, we have initialized two data points located in a two-dimensional space.

2. Calculate the probability associated with the two data points using the `pdf` function, where the mean and variance along the two dimensions is given, as follows:

```
multivariate_normal.pdf(x, mean=[0, 1], cov=[5, 2])
array([ 0.0354664 ,  0.00215671])
```

The output of the preceding code results in probabilities associated with the first data point `[1, 2]` and the second data point `[3, 4]`.

Obtaining random samples of a distribution

Random numbers can be generated using multiple functions within `numpy.random`.

The following table lists the set of functions that are used to generate random data:

Function	Working	Example
Rand	Returns random values in a given shape	`numpy.random.rand(2,2)`
Randn	Returns a sample (or samples) from the standard normal distribution	`numpy.random.randn(2,2)`
Randint	Returns random integers from low (inclusive) to high (exclusive)	`numpy.random.randint(2,20,5)`
Random_integers	Returns random integers of `np.int` type among *low, high,* and *inclusive*	`numpy.random.random_integers(2,20,5)`

| Random | Returns random floats in the half-open interval [0.0, 1.0) | numpy.random.random(5) |

A list of all the possible distributions from which random samples can be drawn is as follows:

Function	Working
beta(a, b[, size])	Draws samples from a beta distribution
binomial(n, p[, size])	Draws samples from a binomial distribution
chisquare(df[, size])	Draws samples from a chi-square distribution
dirichlet(alpha[, size])	Draws samples from the Dirichlet distribution
exponential([scale, size])	Draws samples from an exponential distribution
f(dfnum, dfden[, size])	Draws samples from an F distribution
gamma(shape[, scale, size])	Draws samples from a gamma distribution
geometric(p[, size])	Draws samples from the geometric distribution
gumbel([loc, scale, size])	Draws samples from a Gumbel distribution
hypergeometric(ngood, nbad, nsample[, size])	Draws samples from a hypergeometric distribution
laplace([loc, scale, size])	Draws samples from the Laplace or double exponential distribution with a specified location (or mean) and scale (decay)

`logistic([loc, scale, size])`	Draws samples from a logistic distribution
`lognormal([mean, sigma, size])`	Draws samples from a log-normal distribution
`logseries(p[, size])`	Draws samples from a logarithmic series distribution
`multinomial(n, pvals[, size])`	Draws samples from a multinomial distribution
`multivariate_normal(mean, cov[, size, ...)`	Draws random samples from a multivariate normal distribution
`negative_binomial(n, p[, size])`	Draws samples from a negative binomial distribution
`noncentral_chisquare(df, nonc[, size])`	Draws samples from a noncentral chi-square distribution
`noncentral_f(dfnum, dfden, nonc[, size])`	Draws samples from the noncentral F distribution
`normal([loc, scale, size])`	Draws random samples from a normal (Gaussian) distribution
`pareto(a[, size])`	Draws samples from a Pareto II or Lomax distribution with a specified shape
`poisson([lam, size])`	Draws samples from a Poisson distribution
`power(a[, size])`	Draws samples in [0, 1] from a power distribution with a positive exponent, $a - 1$
`rayleigh([scale, size])`	Draws samples from a Rayleigh distribution
`standard_cauchy([size])`	Draws samples from a standard Cauchy distribution with *mode* = *0*

`standard_exponential([size])`	Draws samples from the standard exponential distribution
`standard_gamma(shape[, size])`	Draws samples from a standard gamma distribution
`standard_normal([size])`	Draws samples from a standard normal distribution (*mean=0, stdev=1*)
`standard_t(df[, size])`	Draws samples from a standard Student's T distribution with *df* degrees of freedom
`triangular(left, mode, right[, size])`	Draws samples from the triangular distribution over the interval [left, right]
`uniform([low, high, size])`	Draws samples from a uniform distribution
`vonmises(mu, kappa[, size])`	Draws samples from a von Mises distribution
`wald(mean, scale[, size])`	Draws samples from a Wald or inverse Gaussian distribution
`weibull(a[, size])`	Draws samples from a Weibull distribution
`zipf(a[, size])`	Draws samples from a Zipf distribution

Given all the preceding functions to generate certain distributions, in order to understand how to generate random samples in a given distribution in Python, we will go through it in the following code:

1. Import the relevant packages:

```
import numpy.random
```

2. Generate 10 random numbers from a standard normal distribution:

```
numpy.random.standard_normal(10)
```

The preceding code generates the following output:

```
array([ 1.19322748,  0.15660068,  0.16968799, -0.9599914 ,
-1.60431604,
        0.31921513, -0.26443506,  1.2290333 ,  1.10691107,
-0.23452172])
```

Getting started with simulation

In order to understand the impact of the number of samples on the distribution of the output obtained, let us look at the following example:

```
import matplotlib.pyplot as plt
plt.hist(numpy.random.standard_normal(10))
plt.show()
```

In the preceding code, we are plotting the histogram distribution of 10 random samples obtained form standard normal distribution.

The output of the code is as follows:

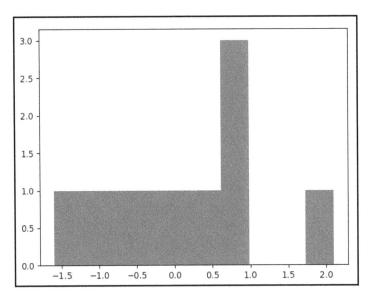

We repeat the preceding code to now generate 100 random samples instead of 10:

```
import matplotlib.pyplot as plt
plt.hist(numpy.random.standard_normal(100))
plt.show()
```

The resulting plot looks like the following:

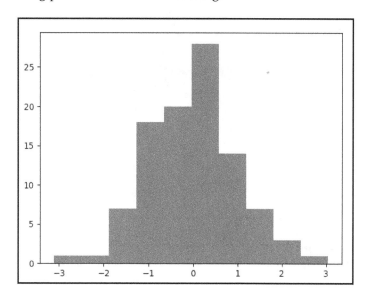

Note that the output of the normal distribution sampling, where we fetch 100 random samples, gives us a lot more normal distribution output than the output when we fetch only 10 random samples from a standard normal distribution.

Furthermore, let's generate a random sample of 1,000 numbers from the normal distribution and compare the results with the previous two outputs:

```
import matplotlib.pyplot as plt
plt.hist(numpy.random.standard_normal(1000))
plt.show()
```

The resulting plot looks like the following:

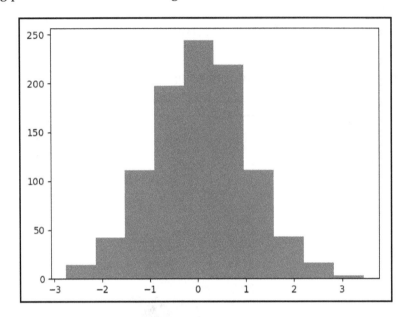

We can see that, as the number of samples fetched increases, the distribution looks more and more like a normal distribution.

Computing the summary statistics of a dataset

Obtaining the summary statistics of a dataset helps us to understand quite a few attributes of the dataset:

- Number of observations in the dataset
- Minimum value and maximum value
- Variance of the dataset
- Mean values in the dataset
- Skewness of the dataset
- Kurtosis of the dataset

How to do it...

The summary statistics of a dataset can be obtained using the `describe` function within `scipy.stats`.

The process to obtain the summary statistics of a dataset is as follows:

1. Import the relevant packages:

   ```
   from scipy import stats
   ```

2. Initialize an array:

   ```
   a = np.arange(10)
   ```

   In the preceding code, we have initialized a one-dimensional array.

3. Fetch the summary statistics of the dataset (array):

   ```
   stats.describe(a)
   DescribeResult(nobs=10, minmax=(0, 9), mean=4.5,
   variance=9.1666666666666661,
                 skewness=0.0, kurtosis=-1.2242424242424244)
   ```

   Note that the result of the `describe` function is all the summary statistics of the dataset.

   The preceding output is for a one-dimensional dataset. In the following example, let us look at calculating the summary statistics of a two-dimensional array (dataset):

1. Initialize a two-dimensional array:

   ```
   b = [[1, 2], [3, 4]]
   ```

2. Calculate the summary statistics of the two-dimensional array:

   ```
   stats.describe(b)
   ```

The output of the preceding line of code is:

```
DescribeResult(nobs=2, minmax=(array([1, 2]), array([3, 4])),
            mean=array([ 2., 3.]), variance=array([ 2., 2.]),
            skewness=array([ 0., 0.]), kurtosis=array([-2.,
-2.]))
```

We can see that the output of `describe` function is all the statistical measures that we discussed earlier in the section.

10
Advanced Computations with SciPy

In this chapter, we will see the following recipes about advanced computing with SciPy:

- Discrete Fourier transforms
- Computing the **discrete Fourier transform** (**DFT**) of a data series using the FFT algorithm
- Computing the inverse DFT of a data series
- Computing signal construction
- Getting started with filters
- Computing the DFT for two-dimensional data
- How to find the DFT of the derivative of a function
- Computing the convolution of two functions
- Mathematical imaging
- Computing pairwise distances from a dataset using different distance metrics
- How to identify neighborhoods and nearest neighbors for a dataset and a given metric
- Nearest neighbors regression

Discrete Fourier transforms

In this section, we will cover the tools surrounding the discrete Fourier and all its statements. A discrete Fourier transform transforms any signal from its time/space domain into a related signal in frequency domain. This allows us not only to analyze the different frequencies of the data, but also enables faster filtering operations, when used properly. It is possible to turn a signal in a frequency domain back to its time/spatial domain, thanks to the **inverse Fourier transform (IFT)**. We will not go into the details of the mathematics behind these operators, since we assume familiarity at some level with this theory. We will focus on syntax and applications instead.

How to do it...

To follow with the example, we need to continue with the following steps:

1. The basic routines in the `scipy.fftpack` module compute the DFT and its inverse, for discrete signals in any dimension—`fft`, `ifft` (one dimension), `fft2`, `ifft2` (two dimensions), and `fftn`, `ifftn` (any number of dimensions).

2. Verify all these routines assume that the data is complex valued. If we know beforehand that a particular dataset is actually real-valued, and should offer real-valued frequencies, we use `rfft` and `irfft` instead, for a faster algorithm.

3. In order to complete with this, these routines are designed so that composition with their inverses always yields the identity.

4. The syntax is the same in all cases, as follows:

```
fft(x[, n, axis, overwrite_x])
```

The first parameter, x, is always the signal in any array-like form. Note that `fft` performs one-dimensional transforms. This means that if x happens to be two-dimensional, for example, `fft` will output another two-dimensional array, where each row is the transform of each row of the original. We can use columns instead, with the optional `axis` parameter. The rest of the parameters are also optional; n indicates the length of the transform and `overwrite_x` gets rid of the original data to save memory and resources. We usually play with the n integer when we need to pad the signal with zeros or truncate it. For a higher dimension, n is substituted by `shape` (a tuple) and `axis` by `axes` (another tuple).

To better understand the output, it is often useful to shift the zero frequencies to the center of the output arrays with fftshift. The inverse of this operation, ifftshift, is also included in the module.

How it works...

The following code shows some of these routines in action when applied to a checkerboard:

```
import numpy
from scipy.fftpack import fft,fft2, fftshift
import matplotlib.pyplot as plt
B=numpy.ones((4,4)); W=numpy.zeros((4,4))
signal = numpy.bmat("B,W;W,B")
onedimfft = fft(signal,n=16)
twodimfft = fft2(signal,shape=(16,16))
plt.figure()
plt.gray()
plt.subplot(121,aspect='equal')
plt.pcolormesh(onedimfft.real)
plt.colorbar(orientation='horizontal')
plt.subplot(122,aspect='equal')
plt.pcolormesh(fftshift(twodimfft.real))
plt.colorbar(orientation='horizontal')
plt.show()
```

Note how the first four rows of the one-dimensional transform are equal (and so are the last four), while the two-dimensional transform (once shifted) presents a peak at the origin and nice symmetries in the frequency domain.

In the following screenshot, which has been obtained from the previous code, the image on the left is the `fft` and the one on the right is the `fft2` of a 2 x 2 checkerboard signal:

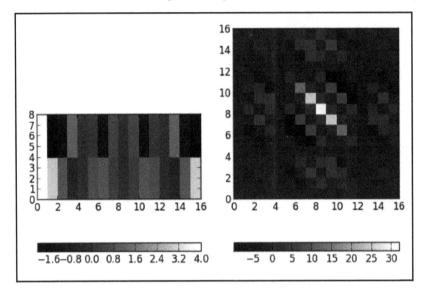

Computing the discrete Fourier transform (DFT) of a data series using the FFT algorithm

In this section, we will see how to compute the discrete Fourier transform and some of its applications.

How to do it...

In the following table, we will see the parameters to create a data series using the FFT algorithm:

Parameters	x: `array_like`. Array to Fourier transform. n: `int`, optional. Length of the Fourier transform. If `n < x.shape[axis]`, x is truncated. If `n > x.shape[axis]`, x is zero-padded. The default results in `n = x.shape[axis]`. axis: `int`, optional. Axis along which the FFTs are computed; the default is over the last axis (that is, axis=-1). `overwrite_x`: Boolean, optional. If `True`, the contents of x can be destroyed; the default is `False`.
Returns	z: complex `ndarray` with the elements: *[y(0),y(1),..,y(n/2),y(1-n/2),...,y(-1)] if n is even* *[y(0),y(1),...,y((n-1)/2),y(-(n-1)/2),...,y(-1)] if n is odd* Where: *y(j) = sum[k=0..n-1] x[k] * exp(-sqrt(-1)*j*k* 2*pi/n), j = 0..n-1* Note that, `y(-j) = y(n-j).conjugate()`.

How it works...

This code represents computing an FFT discrete Fourier in the main part:

```
np.fft.fft(np.exp(2j * np.pi * np.arange(8) / 8))
array([ -3.44505240e-16 +1.14383329e-17j,
         8.00000000e+00 -5.71092652e-15j,
         2.33482938e-16 +1.22460635e-16j,
         1.64863782e-15 +1.77635684e-15j,
         9.95839695e-17 +2.33482938e-16j,
         0.00000000e+00 +1.66837030e-15j,
         1.14383329e-17 +1.22460635e-16j,
        -1.64863782e-15 +1.77635684e-15j])
```

In this example, real input has an FFT that is Hermitian, that is, symmetric in the real part and anti-symmetric in the imaginary part, as described in the `numpy.fft` documentation.

```
import matplotlib.pyplot as plt
t = np.arange(256)
sp = np.fft.fft(np.sin(t))
freq = np.fft.fftfreq(t.shape[-1])
plt.plot(freq, sp.real, freq, sp.imag)
[<matplotlib.lines.Line2D object at 0x...>, <matplotlib.lines.Line2D object
at 0x...>]
plt.show()
```

The following screenshot shows how we represent the results:

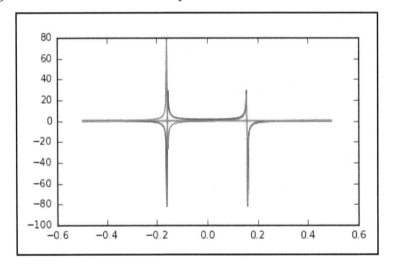

Computing the inverse DFT of a data series

In this section, we will learn how to compute the inverse DFT of a data series.

How to do it...

In this section we will see how to compute the inverse Fourier transform.

The returned complex array contains *y(0), y(1),..., y(n-1)* where:

$$y(j) = (x * exp(2 * pi * sqrt(-1) * j * np.\,arange(n)/n)).\,mean()$$

Parameters	x: `array_like`. Transformed data to invert. n: `int`, optional. Length of the inverse Fourier transform. If n < x.shape[axis], x is truncated. If n > x.shape[axis], x is zero-padded. The default results in n = x.shape[axis]. axis: `int`, optional. Axis along which the IFFTs are computed; the default is over the last axis (that is, axis=-1). overwrite_x: Boolean, optional. If `True`, the contents of x can be destroyed; the default is `False`.
Returns	ifft: `ndarray` of floats. The inverse discrete Fourier transform.

How it works...

In this part, we represent the calculous of the DFT:

```
np.fft.ifft([0, 4, 0, 0])
array([ 1.+0.j,   0.+1.j,  -1.+0.j,   0.-1.j])
```

Create and plot a band-limited signal with random phases:

```
import matplotlib.pyplot as plt
t = np.arange(400)
n = np.zeros((400,), dtype=complex)
n[40:60] = np.exp(1j*np.random.uniform(0, 2*np.pi, (20,)))
s = np.fft.ifft(n)
plt.plot(t, s.real, 'b-', t, s.imag, 'r--')

plt.legend(('real', 'imaginary'))

plt.show()
```

Then we represent it, as shown in the following screenshot:

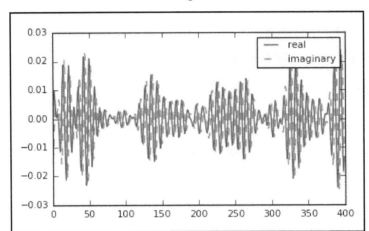

Computing signal construction

In this section, we will see the signal constructions on how to develop the functions.

How to do it...

1. To aid the construction of signals with predetermined properties, the `scipy.signal` module has a nice collection of the most frequent one-dimensional waveforms in the literature—`chirp` and `sweep_poly` (for the frequency-swept cosine generator), `gausspulse` (a Gaussian modulated sinusoid), and `sawtooth` and `square` (for the waveforms with those names).

2. They all take as their main parameter a one-dimensional `ndarray` representing the times at which the signal is to be evaluated. Other parameters control the design of the signal according to frequency or time constraints.

3. Let's take a look into the following code snippet that illustrates the use of these one-dimensional waveforms that we just discussed:

```
import numpy
from scipy.signal import chirp, sawtooth, square, gausspulse
import matplotlib.pyplot as plt
t=numpy.linspace(-1,1,1000)
plt.subplot(221); plt.ylim([-2,2])
plt.plot(t,chirp(t,f0=100,t1=0.5,f1=200)) # plot a chirp
```

```
plt.title("Chirp signal")
plt.subplot(222); plt.ylim([-2,2])
plt.plot(t,gausspulse(t,fc=10,bw=0.5)) # Gauss pulse
plt.title("Gauss pulse")
```

Getting started with filters

In the following section, we will learn how to work with filters.

How to do it...

1. A filter is an operation on signals that either removes features or extracts some component. SciPy has a complete set of known filters as well as the tools to allow the construction of new ones.

2. The complete list of filters in SciPy is long and we encourage the reader to explore the help documents of the `scipy.signal` and `scipy.ndimage` modules for the complete picture.

3. We will introduce in these pages, as an exposition, some of the most used filters in the treatment of audio or image processing.

4. We start by creating a signal worth filtering:

```
from numpy import sin, cos, pi, linspace
f=lambda t: cos(pi*t) + 0.2*sin(5*pi*t+0.1) + 0.2*sin(30*pi*t) +
0.1*sin(32*pi*t+0.1) + 0.1*sin(47* pi*t+0.8)
t=linspace(0,4,400); signal=f(t)
```

First, we test the classical smoothing filter of **Wiener** and **Kolmogorov**, wiener. We present in a plot the original signal (in black) and the corresponding filtered data, with a choice of Wiener window of size 55 samples (in blue). Next, we compare the result of applying the median filter, medfilt, with a kernel of the same size as before (in red).

How it works...

```
from scipy.signal import wiener, medfilt
import matplotlib.pylab as plt
plt.plot(t,signal,'k', label='The signal')
plt.plot(t,wiener(signal,mysize=55),'r',linewidth=3, label='Wiener
filtered')
plt.plot(t,medfilt(signal,kernel_size=55),'b',linewidth=3, label='Medfilt
filtered')
plt.legend()
plt.show()
```

This gives us the following graph showing the comparison of smoothing filters (**Wiener**, in red, is the one that has its starting point just above **0.5** and **Medfilt**, in blue, has its starting point just below **0.5**):

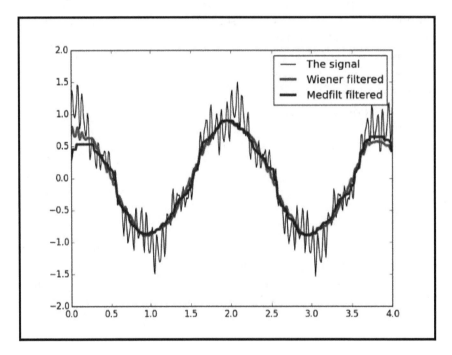

Most of the filters in the `scipy.signal` module can be adapted to work with arrays of any dimension. In the particular case of images, we prefer to use the implementations in the `scipy.ndimage` module, since they are coded with these objects in mind. For instance, to perform a median filter on an image for smoothing, we use `scipy.ndimage.median_filter`. Let us show an example. We will start by loading `lena` to the array, and corrupting the image with Gaussian noise (zero mean and standard deviation of 16):

```
from scipy.stats import norm # Gaussian distribution
import matplotlib.pyplot as plt
import scipy.misc
import scipy.ndimage
plt.gray()
lena=scipy.misc.lena().astype(float)
plt.subplot(221);
plt.imshow(lena)
lena+=norm(loc=0,scale=16).rvs(lena.shape)
plt.subplot(222);
plt.imshow(lena)
denoised_lena = scipy.ndimage.median_filter(lena,3)
plt.subplot(224);
plt.imshow(denoised_lena)
```

The set of filters for images comes in two flavors—statistical and morphological. For example, among the filters of statistical nature, we have the **Sobel** algorithm oriented to the detection of edges (singularities along curves). Its syntax is as follows:

```
sobel(image, axis=-1, output=None, mode='reflect', cval=0.0)
```

The optional parameter, `axis`, indicates the dimension in which the computations are performed. By default, this is always the last axis (-1). The `mode` parameter, which is one of the `reflect`, `constant`, `nearest`, `mirror`, or `wrap` strings, indicates how to handle the border of the image, in the case that there is insufficient data to perform the computations there. In case mode is `constant`, we may indicate the value to use in the border with the `cval` parameter. Let's look into the following code snippet which illustrates the use of the Sobel filter:

```
from scipy.ndimage.filters import sobel
import numpy
lena=scipy.misc.lena()
sblX=sobel(lena,axis=0); sblY=sobel(lena,axis=1)
sbl=numpy.hypot(sblX,sblY)
plt.subplot(223);
plt.imshow(sbl)
plt.show()
```

The following screenshot illustrates the previous two filters in action—Lena (upper-left), noisy Lena (upper-right), edge map with Sobel (lower-left), and median filter (lower-right):

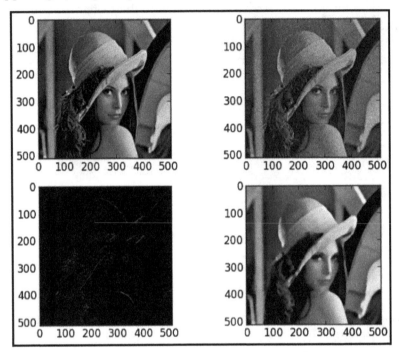

Computing the DFT for two-dimensional data

In this section, we will cover how to compute the DFT for two-dimensional data and its applications.

How to do it...

The following parameters represent the parameter of the functions of the functions for two-dimensional data:

Parameters	x: `array_like`. The (*n*-dimensional) array to transform. shape: Tuple of `int`s, optional. The shape of the result. If both `shape` and `axes` (see as follows) are `None`, shape is x.shape; if `shape` is `None` but `axes` is not `None`, then shape is `scipy.take(x.shape, axes, axis=0)`. If `shape[i]` >x.`shape[i]`, the ith dimension is padded with zeros. If `shape[i]` < x.`shape[i]`, the ith dimension is truncated to length shape[i]. axes: `array_like` of `int`s, optional. The `axes` of x (y if `shape` is not `None`) along which the transform is applied. overwrite_x: Boolean, optional. If `True`, the contents of x can be destroyed. Default is `False`.
Returns	y: Complex-valued *n*-dimensional NumPy array. The (*n*-dimensional) DFT of the input array.

How it works...

This code represents the main calculous of the DFT functions:

```
a = np.mgrid[:5, :5][0]
np.fft.fft2(a)
array([[ 50.0 +0.j        ,    0.0 +0.j        ,    0.0 +0.j         ,
          0.0 +0.j        ,    0.0 +0.j         ],
       [-12.5+17.20477401j,    0.0 +0.j         ,    0.0 +0.j         ,
          0.0 +0.j        ,    0.0 +0.j         ],
       [-12.5 +4.0614962j ,    0.0 +0.j         ,    0.0 +0.j         ,
          0.0 +0.j        ,    0.0 +0.j         ],
       [-12.5 -4.0614962j ,    0.0 +0.j         ,    0.0 +0.j         ,
          0.0 +0.j        ,    0.0 +0.j         ],
       [-12.5-17.20477401j,    0.0 +0.j         ,    0.0 +0.j         ,
          0.0 +0.j        ,    0.0 +0.j         ]])
```

How to find the DFT of the derivative of a function

In this section, we will see how to find the DFT of the derivative of the function.

How to do it...

Return the kth derivative (or integral) of a periodic sequence x.

If x_j and y_j are Fourier coefficients of the periodic functions x and y, respectively, then:

```
y_j = pow(sqrt(-1)*j*2*pi/period, order) * x_j
y_0 = 0 if order is not 0.
```

The previous snippet is just an illustration.

Parameters	x: `array_like`. Input array. order: `int`, optional. The order of differentiation. Default order is 1. If order is negative, then integration is carried out under the assumption that x_0 == 0. period: `float`, optional. The assumed period of the sequence. Default is 2*pi.

Computing the convolution of two functions

In this section, we will cover how to compute the convolution of two functions.

How to do it...

This table represents the parameters of the main functions:

Parameters	x: Input rank-1 `array('d')` with bounds (n). omega: Input rank-1 `array('d')` with bounds (n).
Returns	y: Rank-1 `array('d')` with bounds (n) and x storage.
Other parameters	overwrite_x: Input `int`, optional. Default: 0. swap_real_imag: Input `int`, optional. Default: 0.

Mathematical imaging

In this section, we will cover mathematical imaging using some examples.

How to do it...

First, we need to consider the following points:

- **Mathematical imaging** is a very broad field that is concerned with the treatment of images by representing them as mathematical objects. Depending on the goals, we have four subfields, image acquisition, image compression, image editing, and image analysis.

- **Image acquisition**: The concern here is the effective representation of an object as an image. Clear examples are the digitalization of a photograph (that could be coded as a set of numerical arrays), or super-imposed information of the highest daily temperatures on a map (that could be coded as a discretization of a multivariate function). The processes of acquisition differ depending on what needs to be measured and the hardware that performs the measures. This topic is beyond the scope of this book but, if you are interested, some previous background can be obtained by studying the Python interface to OpenCV and any of the background libraries, such as the **Python Imaging Library** (PIL) and the friendly PIL fork Pillow.

A nice documentation for PIL can be accessed through the `http://effbot.org/` pages at `http://effbot.org/imagingbook/pil-index.htm`. Installing the SciPy stack immediately places a copy of the latest version of PIL in our system. If needed, downloads of this library alone are available from `http://pythonware.com/products/pil/`. For information about Pillow, a good source is `http://pillow.readthedocs.org/`.

A good source of information for OpenCV can be found at `http://opencv.org/`. For a closer look at the interface of Python, I have found the tutorials at `http://docs.opencv.org/3.0-beta/doc/py_tutorials/py_tutorials.html` very useful.

Note that the installation of OpenCV for Python is not easy. My recommendation is to perform such an installation from Anaconda or any other scientific Python distribution.

Computing pairwise distances from a dataset, using different distance metrics

This section represents the pairwise distances from a dataset and some if its applications.

How to do it...

To do this, we need to consider the following points:

- It is imperative to have a good set of different distance functions for any of the algorithms that perform the search and SciPy has, for this purpose, a huge collection of optimally coded functions in the `distance` submodule of the `scipy.spatial` module.
- The list is long. Besides Euclidean, squared Euclidean, or standardized Euclidean, we have many more—Bray-Curtis, Canberra, Chebyshev, Manhattan, correlation distance, cosine distance, dice dissimilarity, Hamming, Jaccard-Needham, Kulsinski, Mahalanobis, and so on.
- The syntax in most cases is simple:

```
distance_function(first_vector, second_vector)
```

The only three cases in which the syntax is different are the Minkowski, Mahalanobis, and standardized Euclidean distances, in which the distance function requires either an integer number (for the order of the norm in the definition of Minkowski distance), a covariance for the Mahalanobis case (but this is an optional requirement), or a variance matrix to standardize the Euclidean distance.

How to identify neighborhoods and nearest neighbors for a dataset and a given metric

This section looks at how to identify neighborhoods and nearest neighbors for a dataset of a given metric and its applications.

How to do it...

This table represents the main parameters of the function to follow:

Parameters	`data: (N,K) array_like`. The data points to be indexed. This array is not copied and so modifying this data will result in bogus results. `leafsize: int`, optional. The number of points at which the algorithm switches over to brute-force. This has to be positive.

Raises	`RuntimeError`: The maximum recursion limit can be exceeded for large datasets. If this happens, either increase the value for the `leafsize` parameter or increase the recursion limit by:
	 `import sys`
	 `sys.setrecursionlimit(10000)`

How it works...

This represents the solution of the main method:

```
import numpy as np
import matplotlib.pyplot as plt
from matplotlib.colors import ListedColormap
from sklearn import neighbors, datasets

n_neighbors = 15

# import some data to play with
iris = datasets.load_iris()

# we only take the first two features. We could avoid this ugly
# slicing by using a two-dim dataset
X = iris.data</span>[:, :2]
y = iris.target

h = .02  # step size in the mesh

# Create color maps
cmap_light = ListedColormap(['#FFAAAA', '#AAFFAA', '#AAAAFF'])
cmap_bold = ListedColormap(['#FF0000', '#00FF00', '#0000FF'])

for weights in ['uniform', 'distance']:
    # we create an instance of Neighbors Classifier and fit the data.
    clf = neighbors.KNeighborsClassifier(n_neighbors, weights=weights)
    clf.fit(X, y)
```

```
# Plot the decision boundary. For that, we will assign a color to each
# point in the mesh [x_min, x_max]x[y_min, y_max].
x_min, x_max = X[:, 0].min() - 1, X[:, 0].max() + 1
y_min, y_max = X[:, 1].min() - 1, X[:, 1].max() + 1
xx, yy = np.meshgrid(np.arange(x_min, x_max, h),
                     np.arange(y_min, y_max, h))
Z = clf.predict(np.c_[xx.ravel(), yy.ravel()])

# Put the result into a color plot
Z = Z.reshape(xx.shape)
plt.figure()
plt.pcolormesh(xx, yy, Z, cmap=cmap_light)

# Plot also the training points
plt.scatter(X[:, 0], X[:, 1], c=y, cmap=cmap_bold,
            edgecolor='k', s=20)
plt.xlim(xx.min(), xx.max())
plt.ylim(yy.min(), yy.max())
plt.title("3-Class classification (k = %i, weights = '%s')"
          % (n_neighbors, weights))

plt.show()
```

This represents the solution of the problem solved:

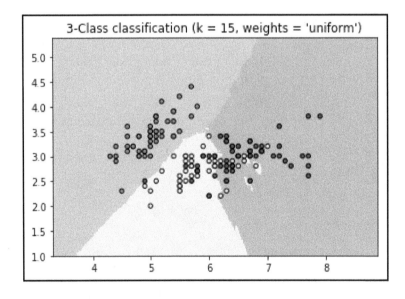

Nearest neighbors regression

This section discusses the nearest neighbors regression.

How it works...

Neighbors-based regression can be used in cases where the data labels are continuous, rather than discrete, variables. The label assigned to a query point is computed based on the mean of the labels of its nearest neighbors.

Scikit-learn implements two different neighbors regressors: KNeighborsRegressor implements learning based on the *K* nearest neighbors of each query point, where *K* is an integer value specified by the user. RadiusNeighborsRegressor implements learning based on the neighbors within a fixed radius *r* of the query point, where *r* is a floating-point value specified by the user.

The basic nearest neighbors regression uses uniform weights. That is, each point in the local neighborhood contributes uniformly to the classification of a query point. Under some circumstances, it can be advantageous to weigh points such that nearby points contribute more to the regression than faraway points. This can be accomplished through the weights keyword. The default value, weights = 'uniform', assigns equal weights to all points. weights = 'distance' assigns weights proportional to the inverse of the distance from the query point. Alternatively, a user-defined function of the distance can be supplied, which will be used to compute the weights.

The code of the Regression:

```
#
############################################################################
##
# Generate sample data
import numpy as np
import matplotlib.pyplot as plt
from sklearn import neighbors

np.random.seed(0)
X = np.sort(5 * np.random.rand(40, 1), axis=0)
T = np.linspace(0, 5, 500)[:, np.newaxis]
y = np.sin(X).ravel()

# Add noise to targets
y[::5] += 1 * (0.5 - np.random.rand(8))
```

```
#
##############################################################################
##
# Fit regression model
n_neighbors = 5

for i, weights in enumerate(['uniform', 'distance']):
    knn = neighbors.KNeighborsRegressor(n_neighbors, weights=weights)
    y_ = knn.fit(X, y).predict(T)

    plt.subplot(2, 1, i + 1)
    plt.scatter(X, y, c='k', label='data')
    plt.plot(T, y_, c='g', label='prediction')
    plt.axis('tight')
    plt.legend()
    plt.title("KNeighborsRegressor (k = %i, weights = '%s')" %
(n_neighbors,
                                                          weights))

plt.show()
```

This represents the final result of nearest neighbor regression:

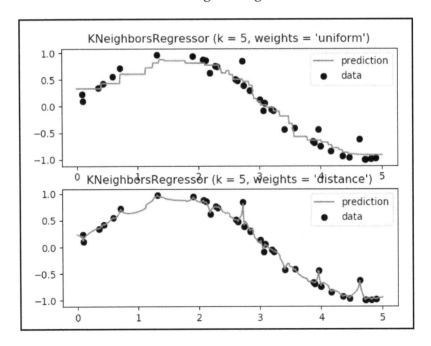

Index

function operating
  defining, on arrays 89, 90, 91
functions
  convolution, computing 352

G

gamma function
  about 255, 258
  working 257
Gaussian quadrature
  used, for computing integrals 270, 272
graphs
  annotating 108, 110, 111
grouping
  performing 176

H

higher moments of distribution
  computing 323, 324
histograms
  generating 112, 113
Homebrew
  installing 15

I

image acquisition 353
indexing 73, 74
integer location
  items, selecting 137
integrals
  computing, Gaussian quadrature used 270, 272
  computing, Newton-Cotes method used 268
  computing, with weighting functions 273
Integrated Development Environment (IDE) 44
integration 266, 267
interactive displays
  generating, in Jupyter Notebook 119, 121
interpolation 276, 277, 278, 279, 281
inverse DFT
  computing, of data series 344
inverse Fourier transform (IFT) 340
items
  selecting, by column labels 135
  selecting, by integer location 137

  selecting, by row indexes 135
  selecting, with mixed indexing 139, 140

J

joining arrays 65, 66
joins
  performing 176
Jordan form of matrix
  calculating 204, 205
Jupyter
  SciPy, executing 38, 39

K

Kolmogorov 347

L

line styles
  setting up 99, 101
linear systems
  solving, matrices used 190, 191
Linux
  SciPy, installing from source 19
lower upper (LU) 197
LU decompositions of matrix
  calculating 197
LU factorization 197

M

macOS
  SciPy, installing from binary distribution 14
map
  creating, with cartopy 124, 127
markers
  reference link 101
  setting up 99, 101
masked arrays
  creating, from condition 86, 87, 88
  creating, from explicit mask 85, 86
  used, to represent invalid data 85
mathematical imaging 353
matrices
  used, for solving linear systems 190, 191
matrix functions
  on two-dimensional arrays 184, 186

P

packages
  installing, with pip 25, 26
pairwise distances
  computing, from dataset with distance metrics
    354
pandas
  configuring 146
Patch API
  reference link 124
PCHIP monotonic cubic interpolation 279
plots
  saving, to disk 105, 106, 107
polynomial interpolation
  computing, for set of data points 282, 283
probabilities associated
  computing, with multivariate Gaussian
    distribution 327, 328, 329, 333
probability density function (pdf)
  about 314
  computing, of continuous random variable 314,
    315, 316, 317, 318
probability mass function (pmf)
  about 308
  binomial discrete distribution 309, 310
  computing, of discrete random variable 308, 309
  multivariate discrete distribution 309, 310, 311,
    312
  visualizing 312, 313, 314
PyCharm
  reference link 44
  SciPy, executing 44, 46, 47, 48, 50, 52
Python 3
  installing 16, 19
Python Imaging Library (PIL)
  about 353
  references 353
Python Package Index (PyPI) 25
Python
  installing 12
  URL 12

Q

QR decomposition of matrix
  calculating 199, 200

R

regression
  about 240
  developing 241, 242
Riemann zeta function 258, 259
row indexes
  items, selecting by 135
rows
  deleting, to DataFrame 134
  inserting, to DataFrame 134
  selecting, of DataFrame 142
  selecting, with Boolean selection 144, 145
Runge-Kutta method 295

S

Scientific PYthon Development EnviRonment
    (Spyder) 40
SciKits
  reference link 23
SciPy stack
  installing 12, 17, 22
SciPy
  executing, in Jupyter 38, 39
  executing, in PyCharm 44, 46, 47, 48, 50, 52
  executing, in script 36
  executing, in Spyder 40, 43
  installing, from binary distribution on macOS 14
  installing, from binary distribution on Windows 11
  installing, from source on Linux 19
  mathematical constants in 246, 247
  physical constants in 246, 247
  reference link 25
series objects
  creating 130, 131
  data, sorting in 175
  numerical functions, applying to 167
  operations, applying to 167
  statistical functions, computing on 169
set of data
  best linear fit, searching 233

www.ingramcontent.com/pod-product-compliance
Lightning Source LLC
Chambersburg PA
CBHW080610060326
40690CB00021B/4644